SRA Real Math

Stephen S. Willoughby

•

Carl Bereiter

•

Peter Hilton

•

Joseph H. Rubinstein

•

Joan Moss

•

Jean Pedersen

Columbus, OH

The **McGraw·Hill** Companies

Authors

Stephen S. Willoughby
Professor Emeritus of Mathematics
University of Arizona
Tucson, AZ

Carl Bereiter
Professor Emeritus
Centre for Applied Cognitive Science
Ontario Institute for Studies in Education
University of Toronto, Canada

Peter Hilton
Distinguished Professor of
Mathematics Emeritus
State University of New York
Binghamton, NY

Joseph H. Rubinstein
Professor of Education
Coker College, Hartsville, SC

Joan Moss
Associate Professor, Department of Human
Development and Applied Psychology
Ontario Institute for Studies in Education
University of Toronto, Canada

Jean Pedersen
Professor, Department of
Mathematics and Computer Science
Santa Clara University, Santa Clara, CA

PreKindergarten and Building Blocks Authors

Douglas H. Clements
Professor of Early Childhood and Mathematics Education
University at Buffalo
State University of New York, NY

Julie Sarama
Associate Professor of Mathematics Education
University at Buffalo
State University of New York, NY

Contributing Authors

Hortensia Soto-Johnson
Assistant Professor of Mathematics
University of Northern Colorado, CO

Erika Walker
Assistant Professor of Mathematics and Education
Teachers College, Columbia University, NY

Research Consultants

Jeremy Kilpatrick
Regents Professor of Mathematics Education
University of Georgia, GA

Alfinio Flores
Professor of Mathematics Education
Arizona State University, AZ

Gilbert J. Cuevas
Professor of Mathematics Education
University of Miami, Coral Gables, FL

Contributing Writers

Holly MacLean, Ed.D., Supervisor Principal, Treasure Valley
Mathematics and Science Center, Boise, ID
Edward Manfre, Mathematics Education Consultant, Albuquerque, NM
Elizabeth Jimenez, English Language Learner Consultant, Pomona, CA

Kim L. Pettig, Ed.D., Instructional Challenge Coordinator
Pittsford Central School District, Pittsford, NY
Rosemary Tolliver, M.Ed., Gifted Coordinator/Curriculum Director, Columbus, OH

National Advisory Board

Justin Anderson, Teacher, Robey Elementary School, Indianapolis, IN
David S. Bradley, Administrator, Granite, UT
Donna M. Bradley, Head of the Lower School, St. Marks Episcopal
Palm Beach Gardens, FL
Grace Dublin, Teacher, Laurelhurst Elementary, Seattle, WA
Leisha W. Fordham, Teacher, Bolton Academy, Atlanta, GA

Ebony Frierson, Teacher, Eastminister Day School, Columbia, SC
Flavia Gunter, Teacher, Morningside Elementary School, Atlanta, GA
Audrey Marie Jacobs, Teacher, Lewis & Clark Elementary, St. Louis, MO
Florencetine Jasmin, Elementary Math Curriculum Specialist, Baltimore, MD
Kim Leitzke, Teacher, Clara Barton Elementary School, Milwaukee, WI
Nick Restivo, Principal, Long Beach High School, Long Island, NY

SRAonline.com

 SRA

Send all inquiries to:
SRA/McGraw-Hill
4400 Easton Commons
Columbus, OH 43219

ISBN 0-07-602998-0

7 8 9 10 11 12 13 RJE 13

Photo Credits

Cover, ©Morgan-Cain & Associates; iii, ©Comstock/
Getty Images, Inc.; iv (l), ©Georgette Douwma/Science
Photo Library/Photo Researchers, Inc, (c), ©Stone/Getty
Images, Inc.

The McGraw·Hill Companies

Number Sense

Exploring 💡 **Problem Solving** Theme: Library

Exploring 💡 Problem Solving Theme: Pets

Exploring 💡 Problem Solving Theme: Origami

Exploring 💡 Problem Solving Theme: Toy Factory

Two-Digit Addition

Exploring 💡Problem Solving   **Theme: Parks and Picnics**

Fractions

CHAPTER 7

Exploring 💡 Problem Solving   Theme: Museums

Geometry

Exploring 💡 **Problem Solving** Theme: Dinosaurs

Three-Digit Addition and Subtraction

Exploring  Problem Solving Theme: Mail Delivery and the Pony Express

CHAPTER 10 Measurement

Exploring Problem Solving Theme: Growing Plants

Introducing Multiplication and Division

Exploring 💡 Problem Solving — Theme: Ethnic Food

CHAPTER 12 Patterns and Algebra

Exploring  Problem Solving Theme: Frontier and Native American Homes

In This Chapter You Will Learn

- counting and writing numbers.
- estimating.
- applications of counting.
- comparing numbers.

Name _____ Date _____

Look at the picture. Try to find all the numbers. Ring each number you find.

1 What does each number tell you?_____

_____

2 How do you know you found all the numbers?

_____

_____

_____

_____

Name _____ Date _____

LESSON 1.1 Counting and Writing Numbers

Key Ideas

You can count on.

| 4 | 5 | 6 |

You can count back.

| 6 | 5 | 4 |

Count on, or count back.
Fill in the missing numbers.

1

5	6			9			12

2

76	77					82	83

3

39	38					33	32

4

63		61				57	56

Number Sequence and Strategies Practice

Tracing and Writing Numbers Game

Players: Two

Materials:

- 0–5 and 5–10 *Number Cubes*

- Different-colored pencil for each player

HOW TO PLAY

❶ Players take turns rolling the 0–5 or the 5–10 *Number Cube* and then tracing or writing the number rolled. Each player uses a different color to write with.

❷ If a player rolls a number that has already been traced or written, the player loses a turn.

❸ When a complete row of numbers has been traced or written, each player counts how many numbers he or she entered.

❹ The player who entered more numbers is the winner.

Tracing Game

| 0 | 1 | 2 | 3 | 4 | 5 | 6 | 7 | 8 | 9 | 10 |

Writing Game

| | | | | | | | | | | |

| | | | | | | | | | | |

Real Math • Chapter 1 • Lesson 1

LESSON 1.2 Odds and Evens

Key Ideas

Even numbers can be split into two equal parts. Odd numbers cannot.

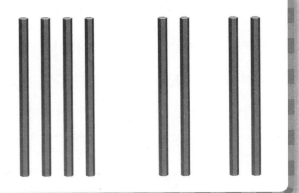

Ring the odd numbers.

Write how many sticks would be in each pile if you split the even numbers.

1 10

2 11

3 12

4 13

5 14

6 15

7 16

8 17

9 18

10 19

11 20

12 21

13 2

14 1

15 4

16 **Extended Response** The house numbers on one side of Franklin Street are 542, 544, and 546.

a. What are the likely house numbers on the other side of the street? _____

b. If you were delivering the mail on Franklin Street, how would you sort it before starting to make deliveries? Explain why. _____

eTextbook This lesson is available in the *eTextbook*.

Game

Number Properties and Sequence Strategies Practice

Odds–Evens Game

Players: Two

Materials: None

HOW TO PLAY

❶ Player One is the even player. Player Two is the odd player.

❷ Each player secretly chooses a number between 0 and 5 and hides one hand behind his or her back with that many fingers showing.

❸ Both players count aloud to 3 together and then bring their hidden fingers to the front.

❹ If the total number of fingers showing is even, the even player wins. If the total number of fingers showing is odd, the odd player wins.

 **Journal**

How many even numbers do you count if you start at 1 and count on by ones to 10? How many odd numbers?

LESSON 1.3 Counting and Estimating

Key Ideas

When you **estimate**, you make a good guess about what you think the answer will be using information you know. If you count the desks in a classroom, you can estimate how many students are in the class.

Complete the estimating activity.

Count Objects	Students	
Estimate Related Objects	Desks	

Count to Check Estimate _____

Count Objects	Shoes	
Estimate Related Objects	Students	

Count to Check Estimate _____

Count Objects		
Estimate Related Objects		

Count to Check Estimate _____

Game

Counting and Writing Numbers Game

Players: Two

Materials: Pencil and paper for each player

HOW TO PLAY

❶ Player One chooses a starting number and an ending number between 0 and 100.

❷ Starting with Player One, players take turns counting, saying, and writing the next one, two, or three numbers. Each player can advance up to three numbers a turn.

❸ Each player must advance at least one number per turn.

❹ The player who says and writes the ending number wins the game.

Sample Game

Start at 27. Stop at 40.

Allison: 27
Geoff: 28, 29

Allison: 30, 31, 32
Geoff: 33

Allison: 34, 35, 36
Geoff: 37

Allison: 38, 39, 40

 **Journal**

If you were playing this game to 40, what number less than 40 would you have to say and write to be sure you would win?

Name _____ Date _____

Making Estimates

Key Ideas

Estimating means using information you have to make good guesses about answers to problems.

Six steps

Twenty steps

About how many steps? _____

About twenty books

About forty books

About how many books? _____

📖 **Textbook** This lesson is available in the *eTextbook.*

The pitcher holds eight glasses of juice.

About how many glasses of juice are in the pitcher? _____

About how many glasses of juice are in the pitcher? _____

Complete the Books to the Top of the Door Activity.

Writing + Math **Journal**

Can you explain why estimating works when you want to find an approximate answer?

LESSON 1.5 Place Value and Money

Key Ideas

When you count 10 or more of an object, you can make a two-digit number by using the **tens place**. The other number stays in the **ones place**.

There are ten sticks in each bundle.

tens place   ones place

5 tens and 7 = 57

You can count money to find out how much you have. Different kinds of money are worth different amounts.

How many? Write your answers.

1  _____

4  _____

2  _____

5 _____

3  _____

6  _____

How much money?

7  _____

8  _____

9  _____

10  _____

11  _____

12  _____

13  _____

14  _____

 Play the **Yard Sale Game.**

Writing + Math **Journal**

How many ways could you make a dollar using only one kind of coin?

Name _____ Date _____

The Calendar

Key Ideas

You can use a calendar to keep track of special days.

December has 31 days.

Fill in the missing numbers.

December

Sunday	Monday	Tuesday	Wednesday	Thursday	Friday	Saturday
			1	2	3	4
5	6	7	8	9	10	11
12	13	14	15	16	17	18
19	20	21	22	23	24	25
26	27	28	29	30	31	

Write the answers.

1. What day is December 5? ___Sun.___
2. What day is December 12? ___Sun.___
3. What day is December 23? ___Thurs.___
4. What day is December 17? ___Fri.___

September has thirty days.

Fill in the missing numbers.

September

Sunday	Monday	Tuesday	Wednesday	Thursday	Friday	Saturday
				1	2	3
4	5	6	7	8	9	10
11	12	13	14	15	16	17
18	19	20	21	22	23	24
25	26	27	28	29	30	

Write the day of the week.

5 September 1 _Thursday_

6 September 2 _Friday_

7 September 14 _Wednesday_

8 September 9 _Friday_

9 What is the twenty-first day of this month? _Wednesday_

10 Saturday, September 3, is the ___3___ day of the month.

11 On September 7, Sam found out that he would have a test on September 14. How much time does Sam have to prepare? _____

 Play the **Calendar Game.**

12 Samir's family is going to visit his aunt for a week. They leave on September 2. While they are there, they decide to stay for an extra two days. What day will they come home?

_____

Writing + Math ⟩ Journal

Do all weeks have the same number of days? Do all months have the same number of days? Do all years have the same number of months?

Name _____ Date _____

Listen to the problem. Think
about how you would solve it.

Atifa decided to use counters to solve
the problem.

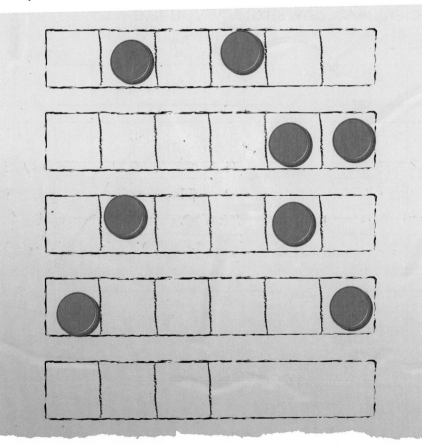

Boyce decided to use a pattern to solve the problem.

What pattern is Boyce using?

Solve the problem. Use any strategy you like.

Mrs. García _____ make a different arrangement each day for ten days.

Cumulative Review

Name _____ Date _____

Numbers on a Clock Grade 1 Lesson 1.11

Write the time below each clock.

1

2

3

_____

_____

_____

Odds and Evens Lesson 1.2

Ring the odd numbers. Write how many would be in each group if you split the even numbers.

4 0

5 31

6 28

7 19

8 10

9 22

Patterns Grade 1 Lesson 1.2

Complete the pattern.

10 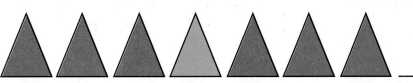 _____

11 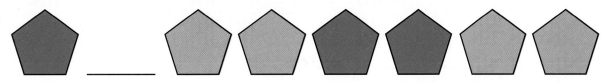

Cumulative Review

Counting and Writing Numbers **Lesson 1.1**

Count on or count back. Fill in the missing numbers.

⑫

42	41					36	35

Making Estimates **Lesson 1.4**

Estimate, and write the number.

This is enough grain for eight piglets.

⑬ About how many
piglets will this feed?

_____

⑭ About how many
piglets will this feed?

_____

Counting by Tens **Grade 1 Lesson 8.6**

Count how many tally marks.
Ring sets of ten to help you count.

||||| ||||| ||||| ||||| ||||| |||||
||||| ||||| ||||| ||||| ||||

⑮ Write the number. _____

⑯ Write the number. _____

Name **Emily** _____ Date _____

Counting on a Number Line

Key Ideas

You can use a number line to count on
and to count back.

What is 18 + 2?

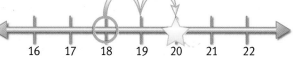

Find 18, and then move on 2 steps.
18 + 2 is 20.

What is 20 − 2?

Find 20, and then move back 2 steps. 20 − 2 is 18.

Draw a ring around the answers.

1 21 + 3

19 20 21 22 23 24 25 26 27 28 29 30

2 21 − 3

15 16 17 18 19 20 21 22 23 24 25 26

3 21 − 2

17 18 19 20 21 22 23 24 25 26 27 28

4 19 + 2

15 16 17 18 19 20 21 22 23 24 25 26

5 0 + 2

0 1 2 3 4 5 6 7 8 9 10 11

6 30 − 2

20 21 22 23 24 25 26 27 28 29 30 31

Find the answers. Watch the signs.

0 1 2 3 4 5 6 7 8 9 10 11 12 13 14 15 16 17 18 19 20 21 22 23 24 25 26 27 28 29 30 31 32 33 34 35

7 $31 + 2 = $ _29_

8 $10 - 1 = $ _9_

9 $30 - 1 = $ _29_

10 $2 + 2 = $ _4_

11 $35 - 2 = $ _33_

12 $7 - 1 = $ _0_

13 $27 - 2 = $ _25_

14 $34 + 1 = $ _35_

15 $9 + 3 = $ _12_

16 $36 - 3 = $ _33_

17 $4 + 1 = $ _5_

18 $32 - 3 = $ _29_

LESSON 1.8 **Counting Applications**

Key Ideas

You can count on and count back to add and to subtract.

Choose the correct answers.

1 53 + 2

51 52 53 54 55 56 57 58 59 60 61 62

2 42 − 3

38 39 40 41 42 43 44 45 46 47 48 49

3 58 + 2

55 56 57 58 59 60 61 62 63 64 65 66

4 15 + 1

12 13 14 15 16 17 18 19 20 21 22 23

5 87 − 2

83 84 85 86 87 88 89 90 91 92 93 94

6 12 − 3

8 9 10 11 12 13 14 15 16 17 18 19

Complete these exercises. Watch the signs.

7 87 + 2 = _____   **8** 88 + 3 = _____   **9** 38 − 3 = _____

10 32 − 3 = _____   **11** 48 + 0 = _____   **12** 11 − 2 = _____

Solve these problems.

⑬ I have $43. If I pay $2 for a library fine, how much money will I have left? _____

⑭ Mr. González is 34 years old.
a. How old will he be two years from now? _____
b. How old will he be three years from now? _____

⑮ James had $89. He bought 2 presents for his sister. Now how much money does he have? _____

⑯ Mahala is 7 years old. She earned $2 today. Now she has $85. How many dollars did she have yesterday? _____

⑰ I have $19. If my aunt gives me $3, how much money will I have? _____

⑱ An airplane holds 67 passengers, 2 pilots, and 1 attendant. How many people does it hold? _____

⑲ Sara had $39. She bought a present for her brother for $2. Then she earned $2 by running errands. How much money does she have now? _____

⑳ **Extended Response** Choose a problem that could not be solved. Tell what information is missing. _____

_____

Writing + Math **Journal**

Create a problem in which you would have to use addition or subtraction to solve. Then write a number sentence to go with your problem.

LESSON 1.9 Comparing Numbers

Key Ideas

The greater-than sign and less-than sign show which amount is greater.

The wider end always points to the greater amount. Use an equal sign if both sides are the same.

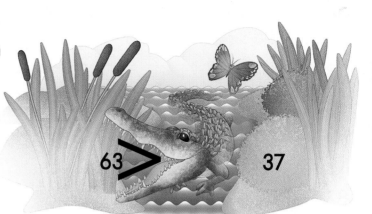

63 > 37

Fill in the blanks on this table.

0	1	2			5	6			9
	11		13	14				18	
20		22				26	27		
	31		33		35		37	38	39
40			43	44	45	46			
50		52		54			57	58	59
	61					66	67	68	69
		72	73	74	75				
80	81	82	83					88	
	92				96			99	
100									

Draw the correct sign: <, >, or =.
Use a number line if you need to.

1 35 ◯ 37

2 45 ◯ 62

3 77 ◯ 82

4 83 + 1 ◯ 83 + 2

5 83 − 1 ◯ 83 − 2

6 83 + 1 ◯ 83 − 1

7 75 + 3 ◯ 75 + 4

8 75 + 23 ◯ 75 + 24

9 12 + 12 ◯ 11 + 13

10 25 + 3 ◯ 25 + 6

11 25 + 4 ◯ 25 + 5

12 25 + 5 ◯ 25 + 4

13 25 + 6 ◯ 25 + 3

14 65 − 3 ◯ 65 − 4

15 66 − 3 ◯ 65 − 2

16 39 + 2 ◯ 39 + 3

17 99 + 0 ◯ 98 + 1

18 87 + 87 ◯ 86 + 86

Real Math • Chapter 1 • Lesson 9

LESSON 1.10

Relating Addition and Subtraction

Key Ideas

Addition and subtraction are inverse operations.

They undo each other.

$$5 + 3 = 8 \qquad 8 - 3 = 5$$

Use the number line to find the answers.

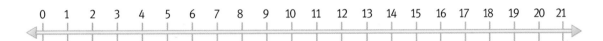

0 1 2 3 4 5 6 7 8 9 10 11 12 13 14 15 16 17 18 19 20 21

1 6 + 3 = ____

2 3 + 1 = ____

3 8 + 2 = ____

4 6 + 1 = ____

5 4 − 1 = ____

6 10 − 2 = ____

7 7 − 1 = ____

8 5 − 2 = ____

9 2 + 3 = ____

10 9 − 3 = ____

11 6 + 2 + 1 + 1 = ____

12 10 − 1 − 1 − 2 = ____

13 2 + 1 + 2 + 1 = ____

14 6 − 1 − 2 − 1 = ____

15 5 + 2 + 2 + 1 = ____

16 10 − 1 − 2 − 2 = ____

Complete the following statements.
Some are easier than they look.

17 5 + 3 = ____

3 + 5 = ____

8 − 5 = ____

8 − 3 = ____

20 13 − 6 = ____

6 + 7 = ____

13 − 7 = ____

7 + 6 = ____

18 5 − 3 = ____

3 + 2 = ____

5 − 2 = ____

2 + 3 = ____

21 3 + 8 = ____

11 − 3 = ____

11 − 8 = ____

8 + 3 = ____

19 9 + 5 = ____

5 + 9 = ____

14 − 5 = ____

14 − 9 = ____

22 234 + 175 = 409

175 + 234 = ____

409 − 175 = ____

409 − 234 = ____

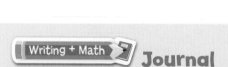 **Journal**

Using the numbers 9 and 2, create as many fact family problems as you can.

Name _____ Date _____

Listen to the problem.

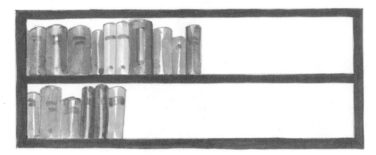

Move _____ books from the top shelf to the bottom shelf.

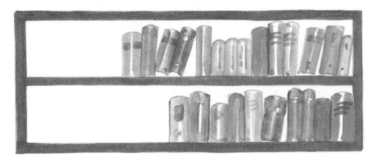

Move _____ books from the top shelf to the bottom shelf.

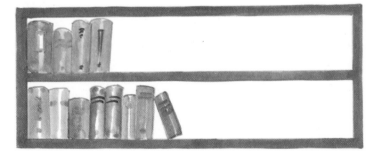

Move _____ books from the bottom shelf to the top shelf.

Listen to the problem. Solve it with a partner.

Move _____ books from one shelf to the other shelf.

Show how you solved the problem.

Cumulative Review

Name _____ Date _____

Finger Sets Grade 1 Lesson 1.6

Write the number shown (number on finger set + number of counters).

1 + = _____

2 + ⬭ ⬭ = _____

3 + ⬭ ⬭ ⬭ = _____

· ·

Place Value and Money Lesson 1.5

Count, and write the amount shown.

4

_____¢

5

$_____

Ⓔ **Textbook** This lesson is available in the *eTextbook*.

Cumulative Review

Comparing Numbers Lesson 1.9

Draw the correct sign: <, >, or =.

6 38 − 1 ◯ 38 − 2

9 82 + 3 ◯ 82 + 2

7 38 + 1 ◯ 38 − 1

10 44 − 1 ◯ 43 − 1

8 67 + 3 ◯ 66 + 4

11 67 − 3 ◯ 66 − 4

Relating Addition and Subtraction Lesson 1.10

Use the number line to find the answers. Some are easier than they look.

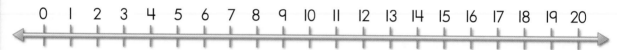

12 5 + 2 = _____

15 5 − 1 − 1 − 2 = _____

13 8 + 1 = _____

16 7 + 3 = _____

14 9 + 1 = _____

17 1 + 2 + 2 + 1 = _____

Solve.

18 234 + 175 = _____

409 − 175 = _____

175 + 234 = _____

409 − 234 = _____

Key Ideas Review

Name _____ Date _____

In this chapter you explored numbers. You learned ways to organize numbers. You compared numbers. You used estimates to solve problems.

· ·

Ring the correct example.

1 tens place

53   53

2 ones place

47   47

3 even number

17   18

4 odd number

16   21

5 The red book has about 100 pages.
The green book has about 500 pages.
About how many pages do you
think the blue book has?

about _____ pages

e Textbook This lesson is available in the *eTextbook*.

6 **Extended Response** Use a picture to show how even numbers are different from odd numbers.

Write the number sentence shown by each number line.

7

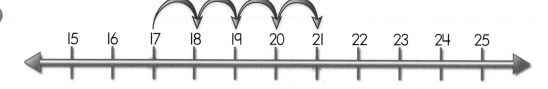

_____

8

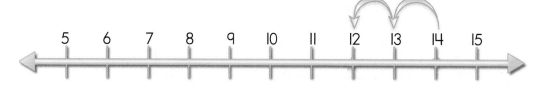

_____

Fill in numbers to make each statement true.

9 $25 + 4 > 25 + \bigcirc$

10 $17 < 15 + \bigcirc$

Name _____ Date _____

Lesson 1.5 **Count,** and then write the number.

1 How many sticks?

_____

2 How much money? _____ dollars

Lesson 1.7 **Solve** these problems using the number line.
Then write the answers.

$$38 \quad 39 \quad 40 \quad 41 \quad 42 \quad 43 \quad 44 \quad 45 \quad 46 \quad 47 \quad 48$$

3 40 + 3 _____

4 39 + 1 _____

Lesson 1.2 **Ring** the odd numbers. Then write how many buttons
would be in each pile if you split the even numbers.

5 38 buttons _____

7 5 buttons _____

6 22 buttons _____

8 35 buttons _____

Lesson 1.4 **Estimate** the number of floors in the first building.

About 10 floors   About 18 floors

9 _____

Lesson 1.1 **Fill** in the missing numbers.

10

58	59			63		65

11

42	41				36	35

Lesson 1.8 **Use** one of the boxes above to help you solve these problems.

12 Mrs. Cho is 62 years old.
   a. How old was she three years ago? _____
   b. How old will she be three years from now? _____

Lesson 1.9 **Draw** the correct sign: <, >, or =.

13 47 ◯ 52   **14** 38 + 1 ◯ 37 + 2

Name _____ Date _____

Write the missing numbers.

1 37, 38, _____, _____, 41

2 16, 17, _____, _____, 20

3 70, _____, _____, 73, 74

Ring two equal groups.

How many craft sticks are in each group?

4 ||||||||||||||| _____

5 ||||||||||||||||||||||||||| _____

Estimate.

6 There are 11 boys on Zach's T-ball team.
The number of girls on the team is about
the same. About how many girls are on
Zach's T-ball team?   About _____

7 Sara is 45 inches tall. Tito is 54 inches tall.
Jong's height is between their heights.
About how tall is Jong?   About _____ inches

Practice Test

How many craft sticks?

8

a. 31
b. 4
c. 30
d. 13

9

a. 45
b. 54
c. 9
d. 50

How much money?

10

a. $6
b. $4
c. $42
d. $24

11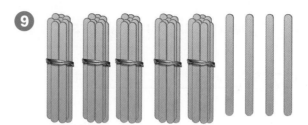

a. $40
b. $45
c. $54
d. $9

Find the answer.

12 50 + 1 = _____

a. 49
b. 50
c. 51
d. 48

13 66 − 2 = _____

a. 65
b. 67
c. 68
d. 64

Name _____ Date _____

Ring the letter of the answer that makes the sentence true.

14 _____ > 37

    a. 35

    b. 39

    c. 33

    d. 37

15 45 < _____

    a. 40

    b. 45

    c. 46

    d. 43

Ring the letter of the answer that shows how many.

16

    a. 10

    b. 14

    c. 12

    d. 11

18

    a. 10

    b. 11

    c. 20

    d. 18

17

    a. 26

    b. 15

    c. 13

    d. 12

19

    a. 18

    b. 17

    c. 20

    d. 19

e Textbook This lesson is available in the *eTextbook*.

Solve.

20 $250 + 125 = 375$

Use this math fact to write one more addition fact and two related subtraction facts.

_____

_____

_____

Use the calendar to solve.

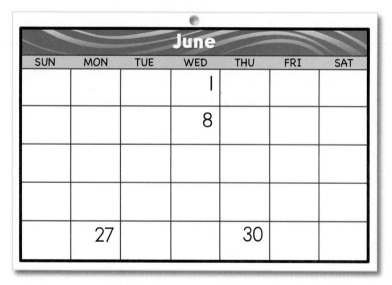

June

SUN	MON	TUE	WED	THU	FRI	SAT
			1			
			8			
	27			30		

21 What is the sixth day of this month? _____

22 How many Wednesdays are in the month? _____

23 What day is June 25? _____

24 Tamara and her dad went fishing on June 19. On what day did they go fishing? _____

Mr. Muddle Takes a Test

Count the number of apples. Cross out 2 of them.
Write the number of apples left.

_____

_____

_____

Draw a line from the number of pigeon eggs to the matching nest.

5

9

12

8

10

Real Math • Chapter 1

Addition Facts

THE ULTIMATE IN CANINE CUISINE !

DELI SPECIAL 299
TURKEY A LA FIDO

DO
DE
S

H²O

Vital 30

CHEESE
YOUR DOG WILL LOVE

ALL NAT

In This Chapter You Will Learn

- basic addition facts.
- about doubles.
- about function machines.

41

Name _____ Date _____

Listen to the problem.

Which bags can you put together to make 16 ounces?

Show as many answers as you can.

Basic Addition Facts and Table

Key Ideas

You can use addition to combine two numbers.
Addition tells how many there are after two sets are combined.

$$5 \quad + \quad 7 \quad = \quad 12$$

The answer to an addition problem is called the sum.

1 Fill in the sums you know on this Addition Table.

+	0	1	2	3	4	5	6	7	8	9	10
0											
1											
2											
3											
4											
5											
6											
7											
8											
9											
10											

Complete the following addition exercises.

2 5 + 1 =   **7** 6 + 2 =

3 2 + 1 =   **8** 6 + 3 =

4 8 + 2 =   **9** 4 + 1 =

5 5 + 2 =   **10** 9 + 0 =

6 7 + 1 =   **11** 3 + 0 =

12 4   **13** 3   **14** 10
    + 3   + 2   + 2

15 8   **16** 7   **17** 8
    + 1   + 3   + 3

18 6   **19** 9   **20** 9
    + 2   + 3   + 2

 Play the **Addition Table Game.**

LESSON 2.3 +10 and +9 Addition Facts

Key Ideas

When you add 10 and a number, the sum is 10 and that number.

$10 + 7 = 17$

The sum of 9 and a number is 1 less than the sum of 10 and that number.

$9 + 7 = 16$

Add to find the sums.

1. $2 + 10 =$

2. $2 + 9 =$

3. $4 + 10 =$

4. $5 + 9 =$

5. $6 + 10 =$

6. $10 + 6 =$

7. $9 + 6 =$

8. $9 + 1 =$

9. $8 + 2 =$

10. $6 + 2 =$

11. $9 + 0 =$

12. $6 + 1 =$

13.
$$\begin{array}{r} 9 \\ + 10 \\ \hline \end{array}$$

14.
$$\begin{array}{r} 9 \\ + 9 \\ \hline \end{array}$$

15.
$$\begin{array}{r} 8 \\ + 10 \\ \hline \end{array}$$

16.
$$\begin{array}{r} 8 \\ + 9 \\ \hline \end{array}$$

e Textbook This lesson is available in the *eTextbook*.

Addition and Strategies Practice

Addition Crossing

Players: Two

Materials:

- *Number Cubes:* two 0–5 (red) and two 5–10 (blue)

- A different color of crayon for each player

HOW TO PLAY

❶ Players roll a 0–5 **Number Cube.** The person who rolls the greater number chooses his or her color of crayon and is followed by the second player.

❷ Take turns rolling any two cubes. Color either square that shows the sum of the addition fact you rolled. (For example, if you roll 3 and 8, you can color the square showing the sum of 3 + 8 or 8 + 3.)

❸ The first player to make a continuous path from one side to the opposite side is the winner. Your path can go up, down, forward, backward, or diagonally as long as all the squares are touching each other.

+	0	1	2	3	4	5	6	7	8	9	10
0	0	1	2	3	4	5	6	7	8	9	10
1	1	2	3	4	5	6	7	8	9	10	11
2	2	3	4	5	6	7	8	9	10	11	12
3	3	4	5	6	7	8	9	10	11	12	13
4	4	5	6	7	8	9	10	11	12	13	14
5	5	6	7	8	9	10	11	12	13	14	15
6	6	7	8	9	10	11	12	13	14	15	16
7	7	8	9	10	11	12	13	14	15	16	17
8	8	9	10	11	12	13	14	15	16	17	18
9	9	10	11	12	13	14	15	16	17	18	19
10	10	11	12	13	14	15	16	17	18	19	20

Name __Emily__   Date _____

LESSON 2.4 **Doubles**

Key Ideas

If you add a number to itself, you double the number. These facts are called the doubles.

Add.

① 8 + 8 = 16   ③ 4 + 4 = 8

② 9 + 9 = 18   ④ 6 + 6 = 12

⑤
```
    5
  + 5
  ----
   10
```
⑥
```
   10
 + 10
 ----
   20
```
⑦
```
    1
  + 1
  ----
    2
```
⑧
```
    0
  + 0
  ----
    0
```

The coats on the Button people have the same number of buttons in back as they do in front. How many buttons are on each coat? Write a number sentence to show how many.

⑨    __2__ + __3__ = __5__

⑩    __3__ + __2__ = __5__

Addition Practice

Doubles Game

Players: Two or three

Materials:

Number Cubes:
two 0–5 (red) and
two 5–10 (blue)

HOW TO PLAY

1 Each player rolls two red and two blue *Number Cubes.*

2 If there are any doubles showing, the player gets 1 point for each sum he or she can correctly name.

3 The first player to get 5 points wins.

 **Journal**

What are the most points you could earn in one turn?
Is it enough to win the game in one turn?

Name Emily   Date _____

Key Ideas

Near doubles are addition facts that are 1 more than a doubles fact. To find $6 + 7$, you can think:
$6 + 6 = 12$, so $6 + 7$ is 1 more than 12.
$6 + 7 = 13$

Add.

1. $6 + 7 = 13$

2. $8 + 7 = 16$

3. $2 + 3 = 5$

4. $5 + 6 = 11$

5. $7 + 8 = 16$

6. $9 + 8 = 17$

7. $4 + 3 = 7$

8. $3 + 2 = 5$

9. $6 + 5 = 11$

10. $2 + 1 = 3$

11. $4 + 5 = 9$

12. $3 + 7 = 10$

13. $7 + 3 = 10$

14. $8 + 1 = 9$

15. $9 + 1 = 10$

16. $7 + 4 = 11$

17.
$$\begin{array}{r} 4 \\ + 7 \\ \hline 11 \end{array}$$

18.
$$\begin{array}{r} 6 \\ + 9 \\ \hline 15 \end{array}$$

19.
$$\begin{array}{r} 5 \\ + 7 \\ \hline 12 \end{array}$$

20.
$$\begin{array}{r} 9 \\ + 6 \\ \hline 15 \end{array}$$

e Textbook This lesson is available in the *eTextbook*.

Answer these questions.

21 Dakota has 5 fish. Ryan has 1 more fish than Dakota. How many fish do they have? _6 fish_

22 Cosmo and Chewie each had 7 dog toys. Then Cosmo lost 1 toy. Now how many toys do they have? _6 dog toys_

23 Josh and Laura each had 5 hermit crabs. Then Laura got 1 more hermit crab. Now how many hermit crabs does Laura have? _6 crabs_

24 Juanita is 8 years old. She bought 8 treat sticks for her canaries. She has used 7 of them. How many treat sticks did Juanita buy? _1 treat_

Name _____ Date _____

Listen to the problem.

Cindy has $3.

She saves $4 each week.

In how many weeks will she have enough money to buy the cage?

Bianca is making a table to solve the problem.

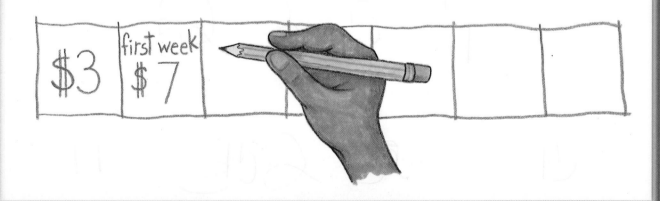

Jorge acts out the problem with play money.

Show how you solved the problem.

Cindy can buy the cage in _____ weeks.

Cumulative Review

Name _____ Date _____

Place Value and Money Lesson 1.5

How much money? Write the answers.

1  _____

2 _____

3 _____

· ·

The Commutative Law Lesson 2.2

Complete these exercises.

4 9 + 2 = ____   **5** 2 + 9 = ____   **6** 1 + 4 = ____

7 4 + 1 = ____   **8** 2 + 6 = ____   **9** 6 + 2 = ____

10   8   **11**   5   **12**   4   **13**   9
     + 5   + 8   + 9   + 4
     ———   ———   ———   ———

· ·

Counting Applications Lesson 1.8

Solve these problems.

14 Katie is 3 centimeters taller than Ann.
Ann is 88 centimeters tall. How tall is Katie? _____

15 Jamal had $80. He bought a present for his
niece for $2. Then he earned $3. How much
money does Jamal have now? _____

Cumulative Review

Comparing Numbers Lesson 1.9

Fill in the blanks on this table.

16

1		3	4			8		
	13		15		17			20
	22	23	24				29	
31				35	36		38	
		44			47		49	
	52		54	55				59	60
	62				66			69	70
71			75			77	78		
81		84	85	86			88		90
91	92				96			99	100

Counting and Writing Numbers Lesson 1.1

Count on or count back.
Fill in the missing numbers.

17

31			29				25	24

18

	13	14			17		

Real Math • Chapter 2

LESSON 2.6

Remaining Facts and Function Machines

Key Ideas

Each function machine has a rule. A number goes into the function machine. The number is changed according to the rule. Then a number comes out of the function machine.

| 3 | in | + 2 | out | 5 |

Find the missing numbers and the rule.

1

in	out
3	5
6	
4	6
	9

The rule is +2.

2

in	out
6	16
4	14
	10
5	

The rule is +10.

3

in	out
4	8
2	4
5	
6	12

The rule is _____.

4

in	out
5	12
8	15
3	
	10

The rule is _____.

5

in	out
2	8
	10
3	9
4	

The rule is _____.

6

in	out
1	10
5	
	11
5	14

The rule is _____.

e Textbook This lesson is available in the *eTextbook*.

Add.

7 0 + 8 =

8 9 + 9 =

9 6 + 6 =

10 3 + 7 =

11 7 + 6 =

12 6 + 8 =

13 9 + 2 =

14 1 + 5 =

15 3 + 6 =

16 10 + 10 =

17 8 + 8 =

18 7 + 2 =

19 7 + 9 =

20 2 + 9 =

21 10 + 4 =

22 9 + 8 =

23
$\begin{array}{r} 6 \\ +\ 9 \\ \hline \end{array}$

24
$\begin{array}{r} 8 \\ +\ 4 \\ \hline \end{array}$

25
$\begin{array}{r} 4 \\ +\ 8 \\ \hline \end{array}$

26
$\begin{array}{r} 9 \\ +\ 6 \\ \hline \end{array}$

LESSON 2.7

Using Multiple Addends

Key Ideas

You can add more than two numbers at once.

$3 + 6 + 7 = 16$

You can find the sum in different ways.

$6 + 7 = 13$	or	$3 + 6 = 9$	or	$3 + 7 = 10$
$13 + 3 = 16$		$9 + 7 = 16$		$10 + 6 = 16$

You might find that one way is easier than another.

Add.

1 $5 + 5 + 4 =$

2 $10 + 4 + 4 =$

3 $10 + 8 + 0 =$

4 $5 + 5 + 5 =$

5 $4 + 8 + 2 =$

6 $1 + 0 + 8 =$

7 $4 + 5 + 9 =$

8 $8 + 2 + 7 =$

9
$$\begin{array}{r} 5 \\ 5 \\ +\ 5 \\ \hline \end{array}$$

10
$$\begin{array}{r} 7 \\ 3 \\ +\ 8 \\ \hline \end{array}$$

11
$$\begin{array}{r} 6 \\ 4 \\ 8 \\ +\ 1 \\ \hline \end{array}$$

12
$$\begin{array}{r} 4 \\ 6 \\ +\ 9 \\ \hline \end{array}$$

13
$$\begin{array}{r} 9 \\ 0 \\ +\ 5 \\ \hline \end{array}$$

14
$$\begin{array}{r} 4 \\ 5 \\ +\ 6 \\ \hline \end{array}$$

15
$$\begin{array}{r} 3 \\ 3 \\ 3 \\ +\ 5 \\ \hline \end{array}$$

16
$$\begin{array}{r} 2 \\ 7 \\ +\ 4 \\ \hline \end{array}$$

Addition and Strategies Practice

Roll a 15 Game

Players: Two

Materials:
Number Cubes:
two 0–5 (red) and
two 5–10 (blue)

HOW TO PLAY

1 Players take turns rolling cubes one at a time. Try to get as close to 15 as possible.

2 Players may stop after any roll. Each cube may be rolled only one time.

3 The player whose sum is closest to 15 wins the round. The number can be greater than or less than 15.

4 The winner of a round rolls first in the next round.

Sample Game

• Palo rolls a red cube. He rolls 5. Then he rolls a blue cube. He rolls 6. His sum is 11. Then he rolls the other red cube. He rolls 2. His sum is 13. Palo stops.

• Gretchen rolls a blue cube. She rolls 7. Then she rolls the other blue cube. She rolls 9. Her sum is 16. Gretchen stops.

• Gretchen wins because her sum is closer to 15.

ⓔGames This game is available as an *eGame.*

Name _____ Date _____

LESSON 2.8 **Applying Addition**

Key Ideas

To find the **perimeter**, add the distances around the outside of a figure.

For example, Mrs. Kinney can add these numbers to find the perimeter of her garden, which is shaped like a triangle:
$3 + 4 + 5 = 12$

3

4

5

Find the perimeter.

①

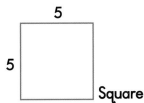

5

5

Square

Perimeter = _____

③

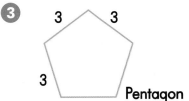

3   3

3

Pentagon

Perimeter = _____

②

6

4

Rectangle

Perimeter = _____

④

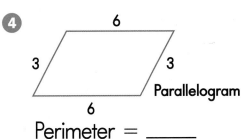

6

3   3

6

Parallelogram

Perimeter = _____

⑤ **Extended Response** Maple Leaf Park is shaped like a rectangle. It has a perimeter of 26 miles.

a. Could two of the sides be 13 miles and 2 miles? _____

b. Why do you think so? _____

ⓔ Textbook This lesson is available in the *eTextbook.*

61

Speed Test

1. $0 + 1 =$ _____

2. $10 + 5 =$ _____

3. $4 + 2 =$ _____

4. $8 + 6 =$ _____

5. $5 + 5 =$ _____

6. $6 + 6 =$ _____

7. $9 + 2 =$ _____

8. $3 + 0 =$ _____

9. $7 + 5 =$ _____

10. $6 + 4 =$ _____

11. $7 + 0 =$ _____

12. $10 + 10 =$ _____

13. $5 + 1 =$ _____

14. $5 + 3 =$ _____

15. $8 + 8 =$ _____

16. $10 + 7 =$ _____

17. $5 + 8 =$ _____

18. $8 + 3 =$ _____

19. $1 + 3 =$ _____

20. $10 + 3 =$ _____

21.
$$\begin{array}{r} 2 \\ + 0 \\ \hline \end{array}$$

22.
$$\begin{array}{r} 5 \\ + 8 \\ \hline \end{array}$$

23.
$$\begin{array}{r} 6 \\ + 9 \\ \hline \end{array}$$

24.
$$\begin{array}{r} 8 \\ + 1 \\ \hline \end{array}$$

25.
$$\begin{array}{r} 3 \\ + 4 \\ \hline \end{array}$$

Name ______________________ Date ______________

LESSON 2.9 The Paper Explorer— Single-Digit Addition

Key Ideas

You can express the same sum in many different ways.

$3 + 5 = 8$ $2 + 6 = 8$ $1 + 7 = 8$

Move counters between the circles. Say the addition facts represented by the groups of counters.

Group 1

Group 2

eTextbook This lesson is available in the ***eTextbook.***

0	1	2	3	4	5	6	7	8	9	10

Add with the Paper Explorer. Try to get sums to match those below.

1. $6 + 5 = 11$

2. $8 + 7 = 15$

3. $7 + 6 = 13$

4. $5 + 9 = 14$

5. $9 + 9 = 18$

6. $8 + 8 = 16$

Name _____ Date _____

Listen to the problem.

Show how to make two baskets of equal value.

Show how to make two baskets of equal value.

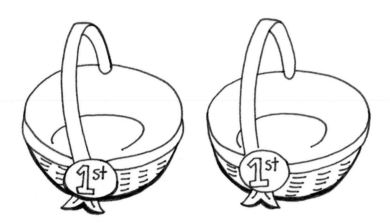

Show how you solved the problem.

Cumulative Review

Name _____ Date _____

Place Value and Money Lesson 1.5

Draw coins to make the correct amount.
Try to use the fewest number of coins possible.

1 65¢

2 14¢

3 23¢

4 37¢

Comparing Numbers Lesson 1.9

Draw the correct sign: <, >, or =.

5 12 + 12 ◯ 12 + 13   **8** 17 + 6 ◯ 19 + 4

6 19 + 5 ◯ 19 + 4   **9** 19 + 4 ◯ 19 + 5

7 19 + 3 ◯ 19 + 6   **10** 79 − 3 ◯ 79 − 4

Cumulative Review

Relating Addition and Subtraction Lesson 1.10

Use the number line to find the answers.

11 6 − 1 = _____

12 7 − 3 = _____

13 3 − 2 = _____

14 4 + 2 + 2 + 1 = _____

15 1 + 3 = _____

16 8 − 1 − 1 − 1 = _____

17 8 − 2 = _____

18 4 + 2 + 1 + 1 = _____

19 6 − 2 = _____

20 7 − 1 − 2 − 1 = _____

```
10
 9
 8
 7
 6
 5
 4
 3
 2
 1
 0
```

Applying Addition Lesson 2.8

Find the perimeter.

21 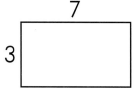 Rectangle

Perimeter = _____

22 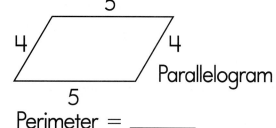 Parallelogram

Perimeter = _____

Add.

23 4 + 5 + 6 = _____

24 4 + 8 + 2 = _____

25 4 + 6 + 9 = _____

26 2 + 7 + 4 = _____

Name _____ Date _____

In this chapter you practiced addition. You learned to use the Addition Table and how to remember addition facts. You used addition to solve problems.

Follow the directions.

1 Color the doubles facts.

2 Color the +10 facts.

+	0	1	2	3	4	5	6	7	8	9	10
0	0	1	2	3	4	5	6	7	8	9	10
1	1	2	3	4	5	6	7	8	9	10	11
2	2	3	4	5	6	7	8	9	10	11	12
3	3	4	5	6	7	8	9	10	11	12	13
4	4	5	6	7	8	9	10	11	12	13	14
5	5	6	7	8	9	10	11	12	13	14	15
6	6	7	8	9	10	11	12	13	14	15	16
7	7	8	9	10	11	12	13	14	15	16	17
8	8	9	10	11	12	13	14	15	16	17	18
9	9	10	11	12	13	14	15	16	17	18	19
10	10	11	12	13	14	15	16	17	18	19	20

3 Fill in this +2 function machine with inputs and outputs that follow the rule.

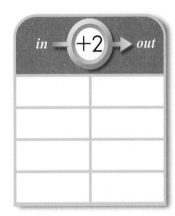

Use words or pictures to answer the following questions.

4 How does knowing the sum of $10 + 4$ help you know the sum of $9 + 4$?

5 How would you find the perimeter of this rectangle?

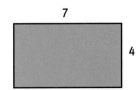

7

4

e Textbook This lesson is available in the *eTextbook*.

Ring the letter of the correct answer.

6 If you know 2 + 6 = 8, then you also know that 6 + 2 = 8. This is an example of _____.

   a. near-doubles facts

   b. doubles facts

   c. the Commutative Law

7 Ethan is adding with the Paper Explorer. He can use the Paper Explorer to show that 8 + 8 is the same as _____.

   a. 9 + 9

   b. 9 + 7

   c. 7 + 7

Write an addition fact you can use to solve each problem.

8 Kori had 5 hamsters. Then she got 1 more. Now how many hamsters does Kori have?

_____ + _____ = _____

9 Jasmine bought 3 fish. Jamal bought 1 less. How many fish did they buy altogether?

_____ + _____ = _____

10 Mr. Thomas needs to buy new horseshoes for his horses. He has 2 horses. Each horse needs 4 shoes. How many horseshoes does Mr. Thomas need to buy?

_____ + _____ = _____

Name _____ Date _____

Lesson 2.7 **Figure** out how much each group of items costs. Write the answers.

1. 3 cat cards _____

2. 1 cat card and 1 horse card _____

3. 1 cat card, 1 horse card, and 1 dog card _____

Lesson 2.3 **Add.**

4. 10 + 8 = _____

5. 9 + 8 = _____

6. 2 + 10 = _____

7. 2 + 9 = _____

8. 5 + 10 = _____

9. 5 + 9 = _____

Lesson 2.6 **Find** the missing numbers and the rule.

10.

in	out
3	11
5	13
	10
6	

The rule is _____.

11.

in	out
3	6
7	14
2	
5	10

The rule is _____.

Lessons 2.1–2.5 ## Add.

12 $2 + 4 =$ _____ **13** $0 + 3 =$ _____ **14** $1 + 7 =$ _____

15 $\begin{array}{r} 4 \\ + 2 \\ \hline \end{array}$ **16** $\begin{array}{r} 3 \\ + 0 \\ \hline \end{array}$ **17** $\begin{array}{r} 7 \\ + 1 \\ \hline \end{array}$ **18** $\begin{array}{r} 6 \\ + 2 \\ \hline \end{array}$ **19** $\begin{array}{r} 2 \\ + 6 \\ \hline \end{array}$

20 $10 + 10 =$ _____ **21** $1 + 1 =$ _____ **22** $3 + 3 =$ _____

23 $8 + 7 =$ _____ **24** $6 + 5 =$ _____ **25** $8 + 9 =$ _____

26 $\begin{array}{r} 7 \\ + 7 \\ \hline \end{array}$ **27** $\begin{array}{r} 6 \\ + 6 \\ \hline \end{array}$ **28** $\begin{array}{r} 9 \\ + 9 \\ \hline \end{array}$ **29** $\begin{array}{r} 9 \\ + 10 \\ \hline \end{array}$ **30** $\begin{array}{r} 6 \\ + 5 \\ \hline \end{array}$

Lesson 2.8 ## Find the perimeter.

31

Pentagon Perimeter = _____

32 **Extended Response** My porch is shaped like a square. It has a perimeter of 20 feet.

a. Could two of the sides be 5 feet and 3 feet? _____

b. Why do you think so? _____

_____

Practice Test

Name _____ Date _____

Add.

1 4 + 4 = _____

3 7 + 7 = _____

2 3 + 4 = _____

4 9 + 9 = _____

5
$$\begin{array}{r} 5 \\ + 6 \\ \hline \end{array}$$

6
$$\begin{array}{r} 6 \\ + 6 \\ \hline \end{array}$$

7
$$\begin{array}{r} 5 \\ + 5 \\ \hline \end{array}$$

8
$$\begin{array}{r} 9 \\ + 8 \\ \hline \end{array}$$

Add. Then use the same numbers to write another addition fact.

9 4 + 6 = _____

10 7 + 5 = _____

_____ + _____ = _____

_____ + _____ = _____

Solve.

11 Maya cut out a square. One side of the square is 3 inches long. What is the perimeter of the square? _____

12 Mike drew a rectangle. Two sides of the rectangle are 4 inches long. The other two sides of the rectangle are 3 inches long. What is the perimeter of the rectangle? _____

Solve.

13 Jean had 7 new pencils. Then she got 8 more new pencils. How many new pencils did she have then?

a. 7　　　**b.** 8　　　**c.** 15　　　**d.** can't tell

14 Jada sees 6 birds on the fence. She sees 4 birds in the tree. How many birds does she see altogether?

a. 2　　　**b.** 4　　　**c.** 6　　　**d.** 10

15 Ashley read 12 pages in her book in the morning. Then she read 10 pages in the afternoon. How many pages did Ashley read altogether?

a. 2　　　**b.** 10　　　**c.** 12　　　**d.** 22

Solve. Use the pictures to find the answers.

16 How much money is it for 2 backpacks?

a. $20　　**b.** $12　　**c.** $10　　**d.** $4

17 How much money is it for 1 skateboard and 1 paint set?

a. $8　　**b.** $10　　**c.** $12　　**d.** $14

18 How much money is it for 1 skateboard, 1 backpack, and 1 paint set?

a. $12　　**b.** $16　　**c.** $18　　**d.** $24

Name _____ Date _____

How much money?

19

   a. $28   **b.** $30   **c.** $38   **d.** $82

20

   a. $4   **b.** $8   **c.** $35   **d.** $53

Choose the number that makes the sentence true.

21 _____ < 71

   a. 90
   b. 87
   c. 77
   d. 70

23 _____ < 29

   a. 31
   b. 29
   c. 22
   d. 92

22 30 > _____

   a. 29
   b. 30
   c. 31
   d. 32

24 43 < _____

   a. 33
   b. 38
   c. 43
   d. 45

Solve.

25 **Extended Response** Write the missing numbers in the Addition Table.

+	0	1	2	3	4	5	6	7	8	9	10
0	0		2			5			8		
1	1			4		6		8	9	10	
2		3	4		6			9	10		
3		4		6				10			13

What is one pattern in the table? _____

_____

_____

26 **Extended Response** **Create** two different function tables. Write the rule for each table.

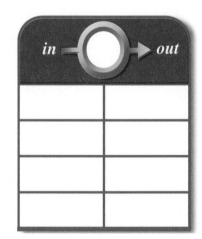

The rule is _____.   The rule is _____.

Name _____   Date _____

Plenty of Time

Untangle the leashes. Trace the path of each leash to the correct dog.

Look at the list of errands. How long will the errands take altogether?

To Do List

Walk dog—   10 minutes

Buy dog food—   15 minutes

          5 minutes

Brush dog—

Give dog a bath—   20 minutes

Total time _____

    Real Math • Chapter 2

In This Chapter You Will Learn

- basic subtraction facts.
- applications of addition and subtraction.
- chain calculations.

Name _____ Date _____

Which is stronger?

Measure to find out. Record your results in the table.

Paper Bridge	Amount It Held

The paper for the stronger bridge held _____ more scissors than the paper for the weaker bridge.

Show how you solved it.

LESSON 3.1 Missing Addends and Subtraction

Key Ideas

Anna needs 8 dollars. She already has 5 dollars.
How many more dollars does she need?

$$5 + \underline{}3\underline{} = 8$$

$$8 - 5 = \underline{}3\underline{}$$

Solve for each missing term.

1 $3 + \underline{} = 8$

2 $8 - 3 = \underline{}$

3 $7 + \underline{} = 14$

4 $14 - 7 = \underline{}$

5 $\underline{} + 5 = 8$

6 $8 - 5 = \underline{}$

7 $4 + \underline{} = 14$

8 $14 - 4 = \underline{}$

9 $9 + \underline{} = 15$

10 $15 - 6 = \underline{}$

11 $\underline{} + 10 = 14$

12 $14 - 4 = \underline{}$

Solve.

⑬ $8 + \underline{\quad} = 15$

⑭ $15 - 7 = \underline{\quad}$

⑮ $8 + \underline{\quad} = 16$

⑯ $16 - 8 = \underline{\quad}$

⑰ $4 + \underline{\quad} = 11$

⑱ $11 - 4 = \underline{\quad}$

⑲ $\underline{\quad} + 9 = 18$

⑳ $18 - 9 = \underline{\quad}$

㉑ $\underline{\quad} + 8 = 11$

㉒ $11 - 3 = \underline{\quad}$

㉓ $10 + \underline{\quad} = 15$

㉔ $15 - 5 = \underline{\quad}$

㉕ **Extended Response** There are 5 swings on the playground. Twelve students want to swing. How many more swings would the playground need for everyone to swing at once? Danny solved the problem this way:

$5 + 7 = 12.$

Melinda solved the problem this way:

$12 - 5 = 7.$

Who is correct?
How would you solve it?

_____

_____

LESSON 3.2 — Subtraction Facts and the Addition Table

Key Ideas

The Addition Table can be used to find subtraction facts.

For every addition fact there is a subtraction fact.

For example:

$2 + 5 = 7$ and $7 - 5 = 2$

Where can you find the subtraction fact $8 - 3 = 5$ on the Addition Table?

You may use the Addition Table to find the answer for each subtraction problem.

1 $17 - 8 =$ _____

2 $13 - 6 =$ _____

3 $6 - 4 =$ _____

4 $9 - 5 =$ _____

5 $15 - 5 =$ _____

6 $11 - 2 =$ _____

7 $9 - 3 =$ _____

8 $16 - 7 =$ _____

+	0	1	2	3	4	5	6	7	8	9	10
0	0	1	2	3	4	5	6	7	8	9	10
1	1	2	3	4	5	6	7	8	9	10	11
2	2	3	4	5	6	7	8	9	10	11	12
3	3	4	5	6	7	8	9	10	11	12	13
4	4	5	6	7	8	9	10	11	12	13	14
5	5	6	7	8	9	10	11	12	13	14	15
6	6	7	8	9	10	11	12	13	14	15	16
7	7	8	9	10	11	12	13	14	15	16	17
8	8	9	10	11	12	13	14	15	16	17	18
9	9	10	11	12	13	14	15	16	17	18	19
10	10	11	12	13	14	15	16	17	18	19	20

e Textbook This lesson is available in the *eTextbook*.

Find the missing numbers.

9 8 + ____ = 17   **11** 8 + ____ = 11

10 11 − 3 = ____   **12** 11 − 8 = ____

13 **Extended Response** If you have $15, can you buy both the chalk and the paint set? Explain your answer.

_____

14 The paint set and scissors cost $15 together. If you have $21, how much money will you have left? ____

15 Two packs of clay cost $12. How much does one pack of clay cost? ____

Writing + Math **Journal**

How would you find the answer to 13 − 4 using the Addition Table?

Name _____ Date _____

LESSON 3.3 Addition and Subtraction Functions

Key Ideas
Function machines can subtract as well as add.

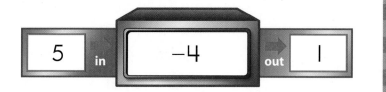

| 5 | in | −4 | out | 1 |

Fill in the missing numbers.

1 in → +5 → out

3	8
5	10
2	

2 in → +6 → out

2	8
	10
3	9

3 in → −3 → out

10	7
8	5
4	

4 in → −7 → out

8	1
9	2
	3
11	

5 in → +3 → out

5	8
10	
9	
1	4

6 in → −4 → out

10	6
	2
6	
5	1

7 in → ○ → out

8	0
18	10
19	
	1

The rule is _____.

8 in → ○ → out

6	12
4	
3	9
	10

The rule is _____.

9 in → ○ → out

30	29
18	17
	34
23	

The rule is _____.

eTextbook This lesson is available in the *eTextbook*.

Game

Subtraction and Strategies Practice

Roll 20 to 5 Game

Players:
Two or three

Materials: Two 0–5 and two 5–10 *Number Cubes* per person

HOW TO PLAY

❶ Player One rolls one of the four cubes and subtracts the number from 20. The player rolls a second cube and subtracts the number from the previous difference. The player may choose to roll a third cube and subtract. The player may choose to do the same with a fourth cube. The player may stop rolling after any of his or her rolls.

❷ Player Two repeats this procedure.

❸ In each round, the player whose score is closer to 5 wins.

❹ The winner of a round rolls first in the next round. In case of a tie, the player who went first in the tie round also goes first in the next round.

e Games This game is available as an *eGame*.

Name _____ Date _____

Listen to the problem.

The Origami Club has 28 sheets of paper.

They use 5 sheets at each meeting.

On what day will they need more paper?

Elena is making a table to help solve the problem.

Day	Mon.	Wed.	Fri.
Number of sheets left	28	23	

Nori is drawing a picture to solve the problem.

Show how you solved the problem.

The club will need more paper on _____ of next week.

Cumulative Review

Name _____ Date _____

The Calendar Lesson 1.6

Fill in the missing numbers. Then use the calendar to answer the questions.

1 Write the dates on the calendar. January has 31 days.

			January			
SUN	MON	TUE	WED	THU	FRI	SAT
1						
			11			
	24					
	31					

Write the answer.

2 What day of the week is January 2? _____

3 What day of the week is January 14? _____

4 What is the fifth day of this month? _____

5 Wednesday, January 4, is the _____ day of the month.

6 On January 9 Josh found out that he would have a test on January 13. How many nights does Josh have to prepare? _____

7 Antwan joined a reading competition that lasts for two weeks. It started Monday, January 2nd. During the competition, Antwan was told they would have 3 extra days. What date will the competition end?

_____

Cumulative Review

Missing Addends and Subtraction Lesson 3.1

Solve.

8 9 + ____ = 17

11 13 − ____ = 5

9 16 − ____ = 13

12 8 + ____ = 11

10 9 + ____ = 15

13 12 − ____ = 6

14 `Extended Response` Explain how to solve the following problem:

____ + 5 = 12. _____

_____

15 There were 16 leaves on the plant. Four were knocked off by the cat. Two more fell off when the dog ran into the plant. The rest stayed on the plant. How many leaves are on the plant? ____

Addition and Subtraction Functions Lesson 3.3

Fill in the blanks.

16

in +11 out	
2	13
4	
3	
5	16

17

in +4 out	
1	5
	9
9	13
8	12

18

in −3 out	
14	11
	14
4	
11	8

LESSON 3.4 Subtraction Involving 10 and 9

Key Ideas

$17 - 10 = 7$
$17 - 9 = 8$

If $18 - 10 = 8$, what is the answer or difference for
$18 - 9?$ _____

Solve the following subtraction problems.

1. $16 - 6 =$ _____

2. $12 - 2 =$ _____

3. $16 - 10 =$ _____

4. $12 - 10 =$ _____

5. $19 - 9 =$ _____

6. $13 - 10 =$ _____

7. $15 - 9 =$ _____

8. $15 - 6 =$ _____

9. $17 - 8 =$ _____

10. $14 - 9 =$ _____

11. $12 - 3 =$ _____

12. $9 - 3 =$ _____

13.
$$\begin{array}{r} 10 \\ -\ 2 \\ \hline \end{array}$$

14.
$$\begin{array}{r} 10 \\ -\ 9 \\ \hline \end{array}$$

15.
$$\begin{array}{r} 10 \\ -\ 4 \\ \hline \end{array}$$

16.
$$\begin{array}{r} 11 \\ -\ 9 \\ \hline \end{array}$$

Answer the following questions.

$7 $6 $2

Kaya has $10.

17 Suppose Kaya buys the ball. How much change will she get? $_____

18 How much change if she buys the car? $_____

19 How much change if she buys the origami paper? $_____

20 **Extended Response** Does Kaya have enough money to buy the ball and the car? _____ Explain your answer. _____

_____

 **Journal**

What are the addition expressions that would go with

$17 - 10 = 7$?

$17 - 9 = 8$?

Game Play the **Roll 20 to 5 Game.**

Name __Emily__   Date _____

Subtraction Facts

Key Ideas

Subtraction is taking away a number of items.

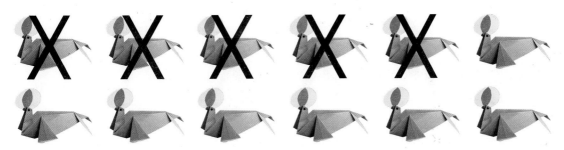

$$12 - 5 = 7$$

Subtract.

1 $16 - 8 =$ __8__

2 $17 - 10 =$ __7__

3 $14 - 5 =$ __10__

4 $15 - 5 =$ __10__

5 $13 - 8 =$ __5__

6 $14 - 7 =$ __7__

7 $10 - 5 =$ __5__

8 $11 - 8 =$ __3__

9 $18 - 9 =$ __9__

10 $9 - 5 =$ __4__

eTextbook This lesson is available in the *eTextbook.*

Speed Test

1 $8 - 8 = \underline{0}$

2 $9 - 7 = \underline{2}$

3 $4 - 2 = \underline{2}$

4 $6 - 5 = \underline{1}$

5 $8 - 4 = \underline{4}$

6 $12 - 9 = \underline{3}$

7 $6 - 1 = \underline{5}$

8 $12 - 8 = \underline{4}$

9 $17 - 5 = \underline{12}$

10 $16 - 6 = \underline{10}$

11 $14 - 7 = \underline{7}$

12 $4 - 1 = \underline{3}$

13 $11 - 7 = \underline{4}$

14 $19 - 6 = \underline{13}$

15 $13 - 9 = \underline{4}$

16 $9 - 4 = \underline{5}$

17 $12 - 3 = \underline{9}$

18 $7 - 6 = \underline{1}$

19 $4 - 4 = \underline{0}$

20 $16 - 10 = \underline{6}$

21
$$\begin{array}{r} 17 \\ -\ 5 \\ \hline 12 \end{array}$$

22
$$\begin{array}{r} 13 \\ -\ 8 \\ \hline 5 \end{array}$$

23
$$\begin{array}{r} 14 \\ -\ 6 \\ \hline 8 \end{array}$$

24
$$\begin{array}{r} 8 \\ -\ 1 \\ \hline 7 \end{array}$$

25
$$\begin{array}{r} 17 \\ -\ 8 \\ \hline 9 \end{array}$$

LESSON 3.6 Applications of Addition and Subtraction

Key Ideas

Addition and subtraction are used to solve problems.

When solving problems, it is important to understand the situation.

Solve.

1 Mark can walk from his house to the library in 6 minutes. About how many minutes will it take him to walk back? _____ About how many minutes will it take him to walk to the library and back? _____ About how many minutes will it take him to walk to school? _____

2 **Extended Response** Sara had $15. She bought a book and now has $9. How much did the book cost? _____ Could she have a $5 bill? Explain. _____ Could she have a $10 bill? Explain. _____

3 **Extended Response** Zane can do a page of addition and subtraction facts in about 10 minutes. His sister can do the same page in about 5 minutes. If they work together, about how many minutes will it take them? Explain. _____ _____

e Textbook This lesson is available in the *eTextbook*.

4 **Extended Response** Max has $3. He wants to buy a ball for $10. He can earn $2 an hour mowing lawns. How many hours must he mow lawns to have $10? Explain. _____

_____

5 **Extended Response** How many bananas can you buy if you have $4 and bananas cost 50¢ each? Explain. _____

_____

6 The score was Sluggers 9 and Bearcats 3 at the end of the eighth inning. If the Sluggers score 1 more run, how many runs must the Bearcats score to win the game? _____

Game Play the **Roll 20 to 5 Game.**

LESSON 3.7

Chain Calculations

Key Ideas

When adding more than two numbers, it is helpful to work the problem in parts.

$5 + 3 + 7 = ?$

First add $5 + 3$.

Then add 7 to that answer.

$$\begin{array}{r} 8 \\ + 7 \\ \hline 15 \end{array}$$

Work these problems. Watch the signs.

1 $11 - 4 + 4 =$ _____

4 $16 - 0 - 8 =$ _____

2 $5 + 5 + 5 =$ _____

5 $4 + 8 - 2 =$ _____

3 $10 - 8 + 7 =$ _____

6 $4 + 5 + 9 =$ _____

7
$$\begin{array}{r} 8 \\ 2 \\ + 7 \\ \hline \end{array}$$

8
$$\begin{array}{r} 5 \\ 5 \\ + 5 \\ \hline \end{array}$$

9
$$\begin{array}{r} 7 \\ 3 \\ + 8 \\ \hline \end{array}$$

10
$$\begin{array}{r} 6 \\ 5 \\ + 8 \\ \hline \end{array}$$

11
$$\begin{array}{r} 6 \\ 6 \\ + 9 \\ \hline \end{array}$$

12
$$\begin{array}{r} 9 \\ 0 \\ + 5 \\ \hline \end{array}$$

13
$$\begin{array}{r} 4 \\ 5 \\ + 6 \\ \hline \end{array}$$

14
$$\begin{array}{r} 3 \\ 3 \\ + 3 \\ \hline \end{array}$$

15
$$\begin{array}{r} 2 \\ 7 \\ + 4 \\ \hline \end{array}$$

Textbook This lesson is available in the *eTextbook*.

Aaron went to a baseball game.
The following items were for sale at the game.

16 How much money would it cost to buy 3 baseballs? _____

17 How much for 2 T-shirts and 1 baseball? _____

18 How much would it cost to buy 4 baseball caps?

_____

19 How much for 2 baseballs and 1 baseball cap?

_____

20 **Extended Response** Jason has $15. Can he buy 2 baseballs and 2 baseball caps? _____ Explain how you found your answer.

_____

_____

 Play the **Roll 20 to 5 Game.**

Exploring Problem Solving

Listen to the problem.

Name _____ Date _____

1  ?

2  −6

3 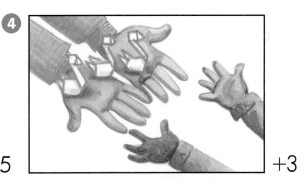 +5

4  +3

5 −7

6 =12

How many origami birds did Joey start with? _____

How do you know? _____

_____

_____

Exploring 💡 Problem Solving

Listen to the problem.

1  ?

2  −4

3 +4

4  +6

5 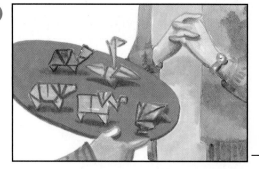 −5

6  =17

How many origami animals did Twyla start with? _____

How do you know? _____

_____

_____

_____

Cumulative Review

Name _____ Date _____

Fractional Parts Grade 1 Lesson 10.7

Ring the fraction that relates to the picture.

1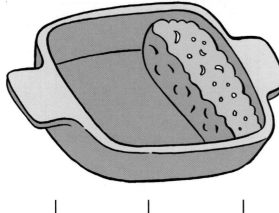

$\frac{1}{2}$ $\frac{1}{3}$ $\frac{1}{4}$

2

$\frac{1}{2}$ $\frac{1}{3}$ $\frac{1}{4}$

Basic Addition Facts and Table Lesson 2.1

Complete the following addition exercises.

3 5 + 3 = _____   **4** 5 + 2 = _____   **5** 5 + 6 = _____

6 2 + 6 = _____   **7** 0 + 1 = _____   **8** 7 + 6 = _____

9 6 + 2 = _____   **10** 1 + 0 = _____   **11** 6 + 7 = _____

12 Sierra is 6 years old. She had a list of 10 chores. She finished 8 of them. How many chores does she have left? _____

Cumulative Review

Making Estimates Lesson 1.4

Estimate and write the number.

⓭ About how many magnets will fit from the bottom to the top of the refrigerator door? (The refrigerator door is on the bottom.) Use the drawings on the refrigerator to help you. _____

Explain. _____

_____

_____

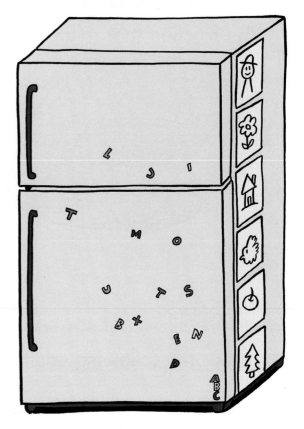

⋯⋯⋯⋯⋯⋯⋯⋯⋯⋯⋯⋯⋯⋯⋯⋯⋯⋯⋯⋯⋯⋯⋯⋯⋯⋯⋯⋯

Chain Calculations Lesson 3.7

Solve.

⓮ $3 + 7 + 8 =$ _____

⓯ $5 + 5 + 5 =$ _____

⓰ $9 + 1 + 3 =$ _____

⓱ $3 + 6 + 8 =$ _____

Name _____ Date _____

In this chapter you learned about subtraction.
You learned about how subtraction is related to
addition. You learned to find answers to problems by
using information you already know.

1 Draw a picture to represent $10 - 5$.

2 Ring the *difference:* $8 - 2 = 6$

Ring the letter of the correct answer.

3 Sami had \$12. She bought a book, and then she
had \$7. Which number sentence can you use to
find out how much the book cost?

   a. $12 + 7$ **b.** $12 - 7$ **c.** $7 - 12$

4 Connor made 4 origami cranes and 3 flowers.
Then he gave away 2 flowers and 2 cranes.
Which number sentence can you use to find out
how many cranes Connor has?

   a. $4 + 3$ **b.** $2 + 2$ **c.** $4 - 3$ **d.** $4 - 2$

5 Knowing $14 - 6 = 8$ also helps you know the
answer to which of the following?

   a. $14 - 8$ **b.** $8 - 6$ **c.** $6 - 14$

6 Solve 4 + 6 + 7. Show your steps.

_____

7 Complete the function table to show the answers to this problem.

Four students each bought a ticket to a movie. A movie ticket costs $4. Maria started with $10. Kyle started with $7. Suzy started with $6. Matt started with $4. How much did each student have after buying a ticket?

8 If 5 + 3 is the same as 3 + 5, then is 5 − 3 the same as 3 − 5? Explain using words or pictures.

_____

_____

9 Maya's class is doing the Numbers on the Back Activity. The sum is 11. What number is on Maya's back? _____

10 What subtraction fact does the table show?

_____

+	0	1	2	3	4	5	6	7	8	9	10
0	0	1	2	3	4	5	6	7	8	9	10
1	1	2	3	4	5	6	7	8	9	10	11
2	2	3	4	5	6	7	8	9	10	11	12
3	3	4	5	6	7	8	9	10	11	12	13
4	4	5	6	7	8	9	10	11	12	13	14
5	5	6	7	8	9	10	11	12	13	14	15
6	6	7	8	9	10	11	12	13	14	15	16
7	7	8	9	10	11	12	13	14	15	16	17
8	8	9	10	11	12	13	14	15	16	17	18
9	9	10	11	12	13	14	15	16	17	18	19
10	10	11	12	13	14	15	16	17	18	19	20

Name _____ Date _____

Lessons 3.2, 3.4, 3.5

Find the difference.

1 $\begin{array}{r} 7 \\ -3 \\ \hline \end{array}$ **2** $\begin{array}{r} 4 \\ -2 \\ \hline \end{array}$ **3** $\begin{array}{r} 9 \\ -4 \\ \hline \end{array}$ **4** $\begin{array}{r} 13 \\ -6 \\ \hline \end{array}$

5 $19 - 4 = $ _____   **6** $17 - 10 = $ _____

Lesson 3.7

Work these problems from left to right.

7 $13 - 3 + 2 = $ _____   **8** $18 - 2 - 2 = $ _____

9 $2 + 2 + 2 = $ _____   **10** $10 + 4 + 3 = $ _____

Lesson 3.6 **11** **Extended Response** Christy has $12 in bills in her purse. Could she have a $5 bill? _____

Explain. _____

50 cents each

12 **Extended Response** How many origami figures can you buy if you have $3 and origami figures cost 50¢ each?

_____

Explain. _____

Lesson 3.3 **Fill** in the missing numbers. Look at the rule.

13 Rule +7

in → +7 → out	
4	11
1	8
6	
3	10

14 Rule −2

in → −2 → out	
9	7
6	4
5	
13	11

Find the rule.

15

in → ◯ → out	
7	2
8	3
19	14
18	13

The rule is _____.

16

in → ◯ → out	
41	39
40	38
	24
15	

The rule is _____.

Lesson 3.1 **Solve** for each missing addend.

17 5 + _____ = 6

18 13 − _____ = 6

19 4 + _____ = 10

20 12 − _____ = 6

21 _____ + 4 = 12

22 13 − _____ = 4

Practice Test

Name _____ Date _____

Solve.

1 Faris had $16. He spent $8 at the carnival. How much money did he have left? _____

2 Paige had one $5 bill and one $10 bill. She wanted to buy a sweater that cost $14. Did she have enough money? Explain. _____

_____

3 Olivia had $12. She bought 2 toys that were $5 each. How much did she have left? _____

Find the rule or the missing numbers.

4

in → +4 → out	
5	9
3	
8	
4	

6

in → ◯ → out	
13	6
14	7
9	2
10	

5

in → −6 → out	
10	4
8	
12	
9	

7

in → ◯ → out	
2	8
9	15
7	13
5	

Practice Test

Ring the letter of the missing number.

8 4 + _____ = 9

    a. 5
    b. 6
    c. 11
    d. 13

10 12 − _____ = 7

    a. 19
    b. 12
    c. 5
    d. 4

9 7 + _____ = 15

    a. 5
    b. 6
    c. 7
    d. 8

11 17 − _____ = 9

    a. 10
    b. 9
    c. 8
    d. 7

Subtract.

12 Carson had $18. He bought a shirt for $9. How much did he have left?

    a. $8
    b. $9
    c. $10
    d. $11

13 Darla had $17. Her sister borrowed $10. How much did Darla have left?

    a. $10
    b. $9
    c. $8
    d. $7

Practice Test

Name _____ Date _____

14 Denzel spent $7 on a pizza and $2 on a drink.
How much change did he get back from a $10 bill?

a. $1   b. $3   c. $8   d. $9

Add or subtract from left to right.

15 3 + 3 + 5 = _____

a. 11   b. 10   c. 8   d. 6

16 5 + 1 + 7 = _____

a. 14   b. 13   c. 12   d. 11

17 16 − 9 − 3 = _____

a. 13   b. 7   c. 5   d. 4

Use doubles to solve.

18 Antonio made 5 base hits in the first baseball game.
He made twice as many base hits in the second game.
How many base hits did he have in the second game?

a. 7   b. 10   c. 12   d. 15

19 Dannah has 6 model cars. Jan has twice as many model
cars as Dannah. How many model cars does Jan have?

a. 6   b. 10   c. 12   d. 14

20 Blake worked 8 hours on Monday and 8 hours on
Tuesday. How many hours did he work on these 2 days?

a. 16   b. 17   c. 18   d. 19

e Textbook This lesson is available in the **eTextbook**.

Extended Response **Solve.**

21 Tia, Dan, and Jacob each had $7. Tia owed Jacob $2. Dan owed Tia $4. Jacob owed Dan $1. Everyone paid what they owed. How much money does each person have now? Explain how you found the amounts.

_____

_____

_____

22 Liliana rode her bicycle from home to school. After school, she rode to a bookstore and to the park. Then she rode home. Use the table to find the total number of miles that Liliana rode her bicycle. Explain how you found the total.

Places	Miles
Home to School	1
Home to Bookstore	4
Home to Park	5
School to Bookstore	3
Park to Bookstore	2

_____

_____

_____

Thinking Story

Ferdie Knows the Rules

Count the number of triangles in the origami animals. Write the number.   _____

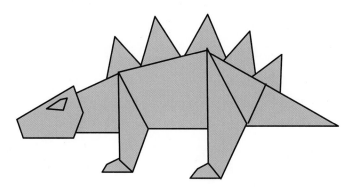

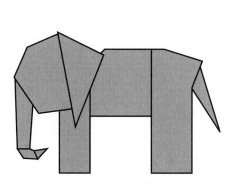

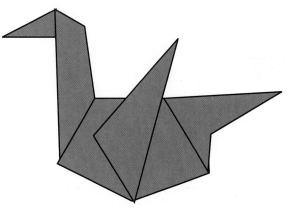

Find a path to the house. Do not get stuck in the mud.

Real Math • Chapter 3

Uses of Measurement, Graphing, and Probability

In This Chapter You Will Learn

- how to measure using standard and nonstandard units.
- how to collect data and display data on graphs.
- about probability.

Name _____ Date _____

Draw your stuffed animal and a hat for it.

Order Form	
How high?	How wide?

Name _____ Date _____

Key Ideas

We can **measure** using objects found in the classroom.

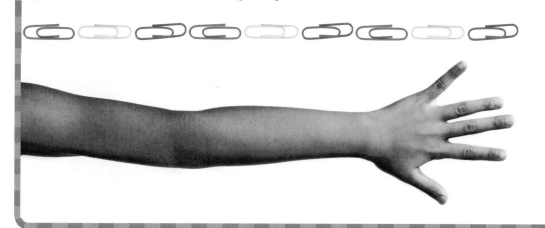

Use paper clips to measure.

1 _____

2 _____

3 **Draw** an outline of your shoe.

4 My body is _____ shoe units long.

> Writing + Math **Journal**
>
> How would you estimate the height of the classroom in shoe units without using a ladder?

LESSON 4.2

Measuring Length—Centimeters

Key Ideas

The centimeter is a unit of **length**.

Estimate the length of each object.
Then measure to check.

1

Estimate _____ cm

Measure _____ cm

2

Estimate _____ cm

Measure _____ cm

3

Estimate _____ cm   Measure _____ cm

Measure the sides. Then add them to find the perimeter.

4 Rectangle

A

D   B

C

Side	Centimeters
A	
B	
C	
D	
Perimeter	

5 Square

A

D   B

C

Side	Centimeters
A	
B	
C	
D	
Perimeter	

6 Triangle

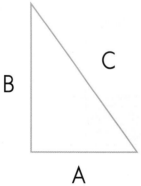

C

B

A

Side	Centimeters
A	
B	
C	
Perimeter	

Game Play the **Find the Distance Game.**

Name _____ Date _____

Measurement—Meters and Centimeters

Key Ideas

**The meter and the centimeter are units of length.
There are 100 centimeters in 1 meter.**

These are some objects that are about
1 meter long.

Do the Measuring Activity.

List three objects in the classroom that are
about 1 meter long. Measure and write their lengths
in centimeters.

Object	Centimeters
1 _____	_____
2 _____	_____
3 _____	_____

e Textbook This lesson is available in the **eTextbook**.

List three objects that are about 2 meters long or 2 meters high.

Measure and write their lengths in centimeters.

Object	Centimeters
4 _____	_____
5 _____	_____
6 _____	_____

I meter = 100 centimeters

Use this information to fill in the blanks.

7 2 m = _____ cm

8 4 m = _____ cm

9 7 m = _____ cm

10 10 m = _____ cm

11 _____ m = 500 cm

12 3 m = _____ cm

13 6 m = _____ cm

14 _____ m = 400 cm

15 _____ m = 800 cm

16 _____ m = 100 cm

LESSON 4.5 Measurement—Yards, Feet, and Inches

Key Ideas

The inch and the foot are units of length. There are 12 inches in 1 foot.

The yard is also a unit of length. There are 3 feet or 36 inches in 1 yard.

We often use these abbreviations:

yard	yd
foot	ft
inch	in.

Fill in the blanks. Use a ruler if you need to.

1 1 foot = _____ inches

2 2 feet = _____ inches

3 1 yard = _____ feet = _____ inches

4 2 yards = _____ feet = _____ inches

Work with a partner to measure.

List some objects in your classroom that are about 1 yard long. Measure them. How many inches long are they?

Object	Nearest Inch
5 _____	_____
6 _____	_____
7 _____	_____
8 _____	_____

Fill in the blanks. Use a ruler if you need to.

9 _____ foot = 12 inches

10 _____ feet = 36 inches

11 _____ feet = 24 inches

12 _____ feet = 60 inches

13 Work with a partner to measure the width of your classroom door. Record the width in inches. _____

14 **Extended Response** How many feet are 48 inches? Explain how you found the answer. _____
_____

Name _____ Date _____

Listen to the problem.

wingspan

Penguin A   Penguin B   Penguin C   Penguin D

Jason
Height | 2 erasers
Wingspan | 3 pennies

Kendra
Height | 2 erasers
Wingspan | 5 pennies

Olivia
Height | 1 paper clip + 1 pen
Wingspan | 5 pennies

Joey
Height | 2 paper clips
Wingspan | 1 eraser + 1 paper clip

Angela is using simple numbers and making a table to solve the problem.

Object	Paper Clip	Eraser	Penny	Tack
Centimeters	3	4		

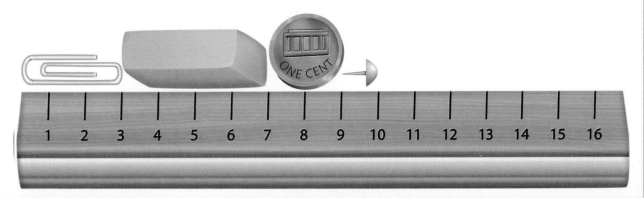

ONE CENT

1 2 3 4 5 6 7 8 9 10 11 12 13 14 15 16

Naresh is making a model to solve the problem.

Solve the problem.

wingspan

Penguin A   Penguin B   Penguin C   Penguin D

_____   _____   _____   _____

Show how you solved the problem.

Cumulative Review

Name _____ Date _____

Counting on a Number Line Lesson 1.7

Draw a ring around the answers.

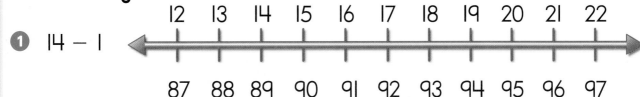

1 14 − 1

2 89 + 0

3 0 + 7

Measuring Length—Centimeters Lesson 4.2

Estimate and then measure in centimeters.

4 Duck estimated length _____
Duck actual length _____

5 Kazoo estimated
length _____
Kazoo actual
length _____

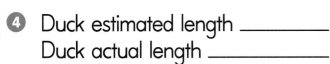

6 Chess piece
estimated height

_____

Chess piece
actual height

_____

7 Which object is
the longest? _____

8 Which is the shortest? _____

_____

Cumulative Review

Place Value and Money Lesson 1.5

How much money?

9   _____

10  _____

....

Missing Addends and Subtraction Lesson 3.1

Fill in the blanks.

11 17 − 9 = _____

12 14 − 9 = _____

13 18 − 9 = _____

14 _____ + 6 = 12

15 18 − 10 = _____

16 _____ + 7 = 14

17 15 − 10 = _____

18 11 + _____ = 17

19 16 − 10 = _____

20 17 − _____ = 9

21 14 − 5 = _____

22 14 − 6 = _____

....

Applying Addition Lesson 2.8

Find the perimeter.

23
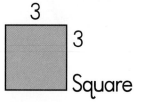
3
3
Square

Perimeter = _____

24
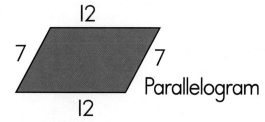
12
7   7
12
Parallelogram

Perimeter = _____

Name ______________________ Date ____________

LESSON 4.6 Collecting and Recording Data

Key Ideas

Tally marks are used to record a number of objects or events.

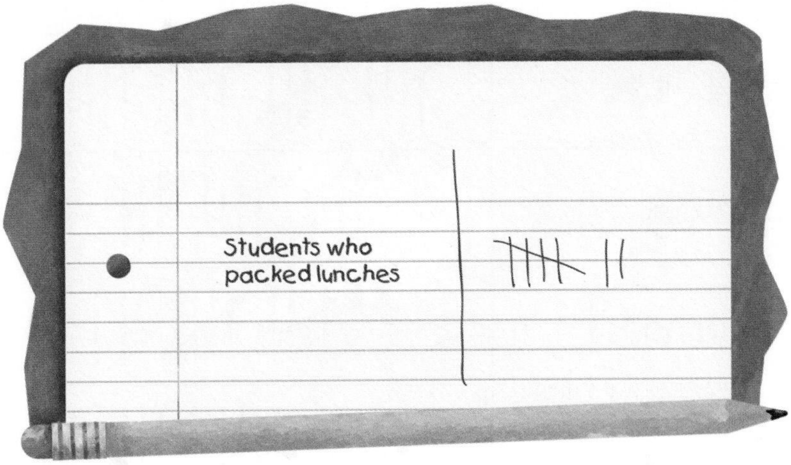
Students who
packed lunches

Which of the letters A, E, I, O, U do you think appears most often in writing?

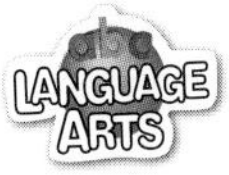
LANGUAGE
ARTS

My prediction ______

My sample ____________________

Letter	Tallies	Totals
A		
E		
I		
O		
U		

Sara made a survey of the 65 seniors at Stuyvesant High School. She asked each senior which sports he or she played. Here are her results.

Sport	How Many
Baseball	~~IIII~~ ~~IIII~~ ~~IIII~~ ~~IIII~~ ~~IIII~~
Volleyball	~~IIII~~ ~~IIII~~ ~~IIII~~ II
Basketball	~~IIII~~ ~~IIII~~ III
Soccer	~~IIII~~ ~~IIII~~ ~~IIII~~ III
Tennis	~~IIII~~ II
Other Sports	IIII

Use the table to answer these questions.

❶ How many students play baseball? _____

❷ How many students play soccer? _____

❸ How many students do not play any sport? _____

❹ **Extended Response** If there are only 65 seniors, how can these numbers be accurate?

_____

130

Name _____ Date _____

LESSON 4.7 Venn Diagrams

Key Ideas

A Venn diagram is a tool for gathering
and comparing data.

1. Complete the Venn diagram with your interests,
   your partner's interests, and interests you share.

About Me   About _____

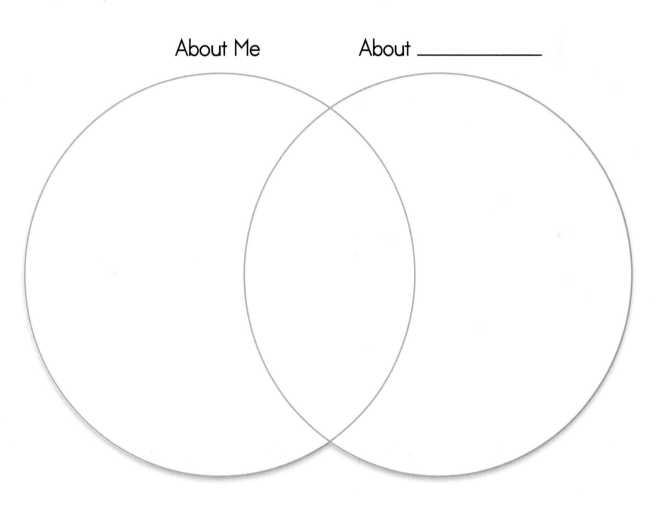

2 Draw lines to show where each figure should go in the Venn diagram.

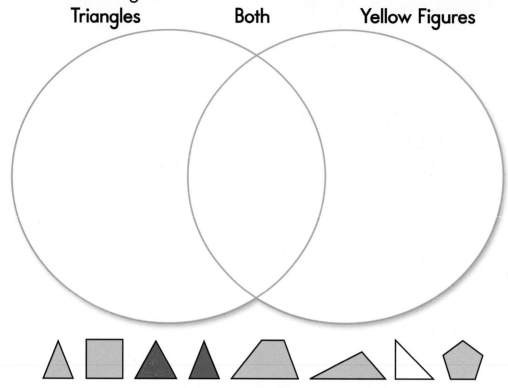

Triangles   **Both**   **Yellow Figures**

3 Rewrite each word in the correct section of the Venn diagram.

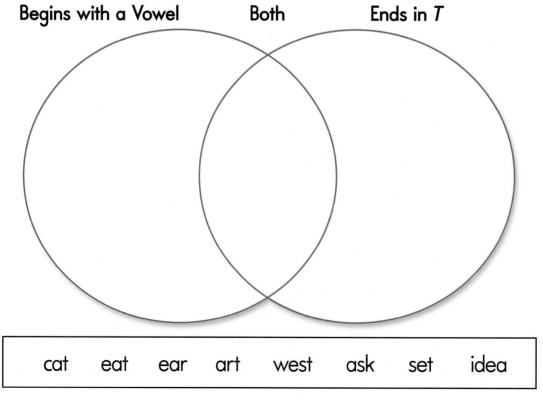

Begins with a Vowel   **Both**   **Ends in _T_**

| cat | eat | ear | art | west | ask | set | idea |

Real Math • Chapter 4 • Lesson 7

Name _____ Date _____

Key Ideas

Mode is the number that appears most often in a set of data. For example, in counting the number of seeds in pods, the number that appears most often is the mode.

The least and greatest number of seeds found in a group of pods is the range.

How many peas are in a pod?

Number of Seeds	How many?	
	Tallies	Number
0		
1		
2		
3		
4		
5		
6		
7		
8		
9		
10		

1 My mode is _____ .   **2** My range is _____ to _____ .

How many peas are in a pod?

Number of Seeds	Individual Reports	Class Totals
0		
1		
2		
3		
4		
5		
6		
7		
8		
9		
10		

3 The class mode is _____.

4 The class range is _____ to _____.

5 **Extended Response** If you examined another batch of the same kind of peas, what do you think the mode would be? _____ Why? _____

LESSON 4.9 Pictographs

Key Ideas

A pictograph is a graph that uses pictures to represent information.

Look at the pictograph below.

Each picture of a can stands for a group of 10 canned goods.

Ms. Allen's Class					
Ms. Beck's Class					
Mr. Carl's Class					

Use the pictograph to answer these questions.

1. Which class collected the greatest number of canned goods? _____ How many? _____

2. How many canned goods did Ms. Allen's class collect? _____

3. How many more canned goods were collected by Mr. Carl's class than by Ms. Beck's class? _____

4. How many canned goods were collected by Ms. Allen's class and Ms. Beck's class altogether? _____

Look at the pictograph below. What does each picture represent?

Each _____ stands for 10 gallons of gas used on a trip between two towns.

Albright to Trent	🛢	🛢	🛢	🛢	🛢	🛢	🛢
Albright to Wayne	🛢	🛢	🛢	🛢			
Trent to Newtown	🛢	🛢	🛢	🛢	🛢		
Newtown to Wayne	🛢	🛢	🛢	🛢	🛢		

Use the pictograph to answer these questions.

5 A trip between which two towns uses the least amount of gas? _____

6 How many gallons are used on a trip from Trent to Newtown? _____

7 A trip between which two towns uses the most gas? _____

8 Extended Response Draw a map that shows where Albright, Trent, Wayne, and Newtown might be. Compare your map with others. Explain how you made your map.

_____

_____

LESSON 4.10

Vertical Bar Graphs

Key Ideas

Bar graphs use shaded areas or bars to represent information. Look at the bar graph below.

What does each bar stand for?

Jenni, Ryan, Ladonna, Daniel, and Adam made a bar graph showing how many books they read in one month.

Use the bar graph to answer these questions.

1 Who has read the most books? _____ How many? _____

2 Who has read the fewest books? _____ How many? _____

3 How many books has Jenni read? _____

4 How many books has Adam read? _____

5 How many more books did Ladonna read than Adam? _____

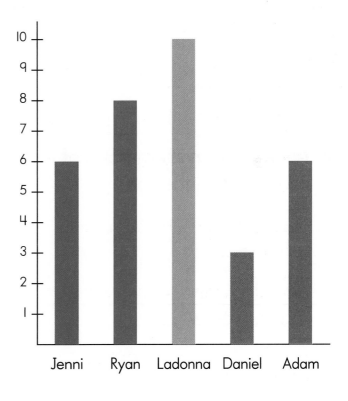

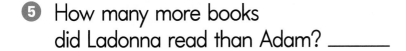 **Textbook** This lesson is available in the *eTextbook*.

These graphs show the results of probability experiments like the one you did in class. For each experiment 18 marbles were in the can.

6 About how many marbles of each color were in the can?

_____ red

_____ blue

_____ yellow

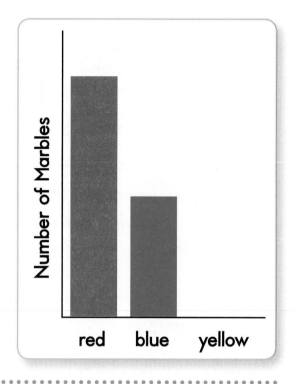

7 About how many marbles of each color were in the can?

_____ red

_____ blue

_____ yellow

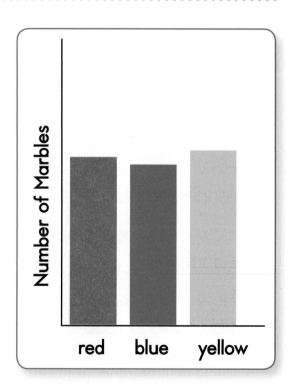

Name _____ Date _____

Horizontal Bar Graphs

Key Ideas

This bar graph shows the average high temperature for each month in a United States city.

Use the bar graph to answer these questions.

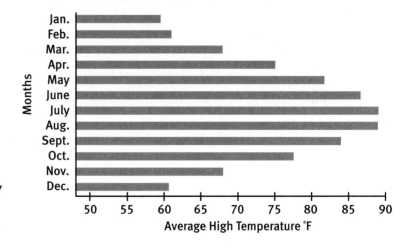

1 What is the average high temperature for April? _____°F

2 Which month is usually the coldest in this city?

_____

3 Which month is usually the warmest? _____

4 **Extended Response** Do you think the temperature ever gets to 100°F in this city? Explain.

_____

_____

5 Gather your own data by doing research. Find the average high temperatures for your city or for a city near you for a year. Use this table to record your data.

Month	Average High Temperature °F
January	
February	
March	
April	
May	
June	
July	
August	
September	
October	
November	
December	

Average High Temperatures for the city of _____

6 Transfer the data you collected to the bar graph.

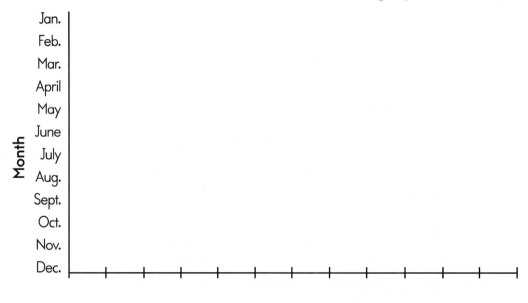

Average High Temperature °F

Name _____ Date _____

Key Ideas

Information can be recorded on a grid.

This is a graph on a grid. What information do you
see on this graph?

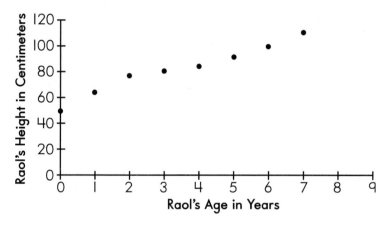

Raol's Height in Centimeters / Raol's Age in Years

Use the graph to answer these questions.

1 Connect the points on the graph.

2 About how tall was Raol when
he was 3 years old? _____ cm

3 About how old was Raol when he
was 100 cm tall? _____

4 About how old was Raol when he
was 70 centimeters tall? _____

5 **Extended Response** About how tall
will Raol be when he is 8 years
old? Explain how you know.

_____

e Textbook This lesson is available in the *eTextbook*.

Max did an experiment to see how long it took for 100 radish seeds to sprout. He planted the seeds and then counted the number that sprouted as of the end of each day. He recorded his results on the following table.

Days	Number Sprouted as of the End of That Day
1	0
2	0
3	5
4	19
5	45
6	70
7	84
8	85
9	85
10	85

Use the table to answer these questions.

6 How many seeds had sprouted by the end of the fifth day? _____

7 On which day did the most seeds sprout? _____

8 How many seeds sprouted on that day?

_____

9 Extended Response Do you think all 100 seeds will sprout? Write why or why not. _____

_____

LESSON 4.13 Using Probability

Key Ideas
Some chance events are more likely than others.

Of the numbers 0–5, which number do you think will appear most often as you roll the cube?

My prediction _____

Do the Cube-Rolling Activity.

❶ Roll a 0–5 cube many times.

❷ Keep track of the numbers you roll.

	Tallies	Totals	Class Totals
0			
1			
2			
3			
4			
5			

ⓔ **Textbook** This lesson is available in the *eTextbook.*

Roll one 0–5 *Number Cube* and one 5–10 *Number Cube.* **Find the sum.**

What sum do you predict will appear most often?

My Prediction _____

Do the Cube-Rolling Activity.
Keep track of the numbers you roll.

Sum	Tallies	Totals	Class Totals
5			
6			
7			
8			
9			
10			
11			
12			
13			
14			
15			

3 **Extended Response** Which sums appeared most often? Which sums appeared least often? Explain why you think they appeared that way. _____

_____

Writing + Math **Journal**

How many different ways could you roll a sum of 5?
How many different ways could you roll a sum of 10?

Exploring Problem Solving

Name _____ Date _____

Listen to the problem. What sizes should the factory make?

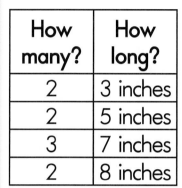

How many?	How long?
2	3 inches
2	5 inches
3	7 inches
2	8 inches

How many?	How long?
2	3 inches
1	5 inches
1	8 inches
1	11 inches

How many?	How long?
1	4 inches
2	6 inches
2	7 inches
3	11 inches

How many?	How long?
2	3 inches
1	9 inches
2	10 inches
1	11 inches

How many?	How long?
1	4 inches
3	6 inches
2	11 inches
1	13 inches

Organize this information.

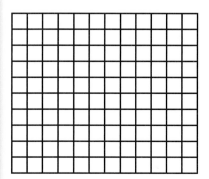

These are the three sizes the factory should make: _____

Exploring Problem Solving

What sizes should the factory make?

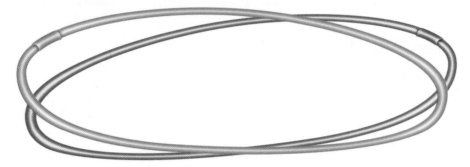

How many?	How long?
2	8 inches
1	10 inches
3	15 inches
1	18 inches

How many?	How long?
1	9 inches
3	11 inches
2	15 inches

How many?	How long?
2	14 inches
1	16 inches
3	18 inches
3	20 inches

How many?	How long?
2	8 inches
3	9 inches
2	12 inches
1	14 inches

How many?	How long?
1	8 inches
3	9 inches
2	15 inches
2	20 inches

Organize this information.

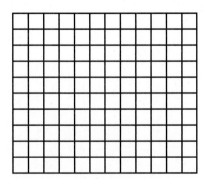

These are the three sizes the factory should make: _____

Cumulative Review

Name _____ Date _____

Copyright © SRA/McGraw-Hill.

Measurement—Meters and Centimeters Lesson 4.3

Fill in the blanks.

1 5 m = _____ cm

4 _____ m = 100 cm

2 10 m = _____ cm

5 _____ m = 800 cm

3 _____ m = 400 cm

6 _____ m = 400 cm

Near Doubles Lesson 2.5

Write the answer.

7 Normie and Tiger each had 6 cat toys. Then Tiger lost 1 toy. Now how many toys do they have altogether? _____

Counting Applications Lesson 1.8

Write the answer.

8 Kalil earned $3 today. Now he has $77. How many dollars did he have yesterday? _____

Chain Calculations Lessons 3.7

Solve.

9
$$
\begin{array}{r} 6 \\ 4 \\ + 3 \\ \hline \end{array}
$$

10
$$
\begin{array}{r} 1 \\ 0 \\ + 8 \\ \hline \end{array}
$$

11
$$
\begin{array}{r} 3 \\ 6 \\ + 4 \\ \hline \end{array}
$$

12
$$
\begin{array}{r} 10 \\ 8 \\ + 0 \\ \hline \end{array}
$$

Cumulative Review

Measurement—Yards, Feet, and Inches **Lesson 4.5**

Fill in the blanks.

⑬ 1 foot = 12 inches

⑮ _____ feet = 48 inches

⑭ _____ feet = 72 inches

⑯ _____ feet = 60 inches

Remaining Facts and Function Machines **Lesson 2.6**

Add.

⑰ 9 + 9 = _____

⑳ 7 + 2 = _____

⑱ 3 + 6 = _____

㉑ 8 + 8 = _____

⑲ 6 + 9 = _____

㉒ 2 + 9 = _____

Applications of Addition and Subtraction **Lesson 3.6**

㉓ **Extended Response** Miguel has $5. He wants to buy a pogo stick for $13. He can earn $2 an hour helping his older sister with her paper route. How many hours must he help with the paper route to have $13? _____ Explain. _____

_____

_____

+10 and +9 Addition Facts **Lesson 2.3**

Add to find the sums.

㉔ 9 + 6 = _____

㉕ 10 + 3 = _____

㉖ 9 + 0 = _____

In this chapter you learned about uses of measurement, graphing, and probability. You learned to measure length in nonstandard and standard units. You learned how to collect, record, and analyze data.

1 Cory wants to measure the length of his desk. He could measure with pencil units or paper-clip units. Which unit would he need more of?

_____

Why? _____

_____

2 _____

_____

Juana measured the red line and found that it is 10 centimeters long. Is 9 centimeters a good estimate for the length of the orange line?_____ Explain why. _____

_____

Ring the letter of the correct answer.

3 1 meter = _____ centimeters

   a. 10   **b.** 100   **c.** 1,000

④ Match the length on the left with the picture on the right that is about that length.

about 1 inch

about 1 yard

about 1 foot

⑤ Draw an example of tally marks.

The graph shows how many books each student read.

⑥ Who read the most books? _____

⑦ How many? _____

⑧ What is the mode?

_____

⑨ What is the range?

_____ to _____

⑩ Anna read 7 books. Add her data to the graph.

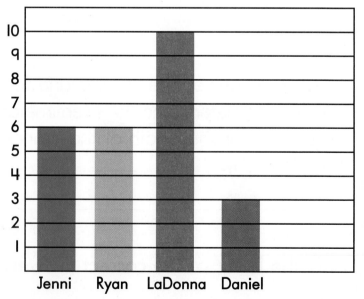

Name _____ Date _____

Lesson 4.10 Use the bar graph to answer these questions.

Number of Books Read

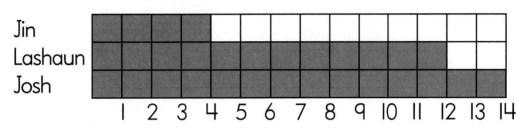

Jin
Lashaun
Josh

1 2 3 4 5 6 7 8 9 10 11 12 13 14

1 Who has read the most books? _____
How many? _____

2 How many books has Lashaun read? _____

Lessons 4.2 and 4.4 How long? Estimate and then measure.

3

Estimate
in centimeters _____

Actual length
_____ centimeters

4

Estimate
in inches _____

Actual length
_____ inches

Lessons 4.6 and 4.8 **Write** the number of tally marks. Then use the table to answer the questions below.

Shantrell made this table. She kept track of the number of swimming pool visitors each hour she was there.

Time of Day	Tallies	Number
10 A.M.	||	
11 A.M.	卌 ||||	
12 P.M.	卌 卌 ||||	
1 P.M.	卌 卌 卌 ||	
2 P.M.	卌 卌 卌 |	
3 P.M.	卌 卌 卌 |	

5 The mode is _____ .

6 The range is _____ to _____ .

Lessons 4.3 and 4.5 **Fill** in the blanks. Remember, 1 meter = 100 centimeters; 1 foot = 12 inches; 1 yard = 3 feet.

7 5 m = _____ cm

8 8 m = _____ cm

9 1 yard = _____ inches

10 2 feet = _____ inches

11 2 yards = _____ feet = _____ inches

Practice Test

Name _____ Date _____

Use the Venn diagram to answer the questions.

Manuel asked his friends what movies they like.

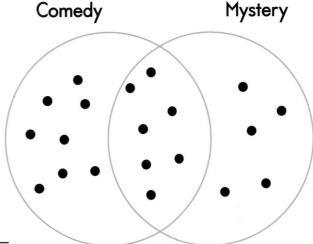

1. How many friends like comedies? _____

2. How many friends like mysteries? _____

3. How many friends like both comedies and mysteries? _____

4. How many friends did Manuel ask? _____

Use the bar graph to answer the questions.

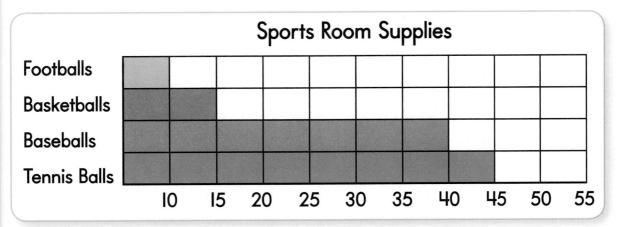

5. What kind of balls are there the fewest of in the sports supply room? _____

6. How many more baseballs are there than footballs? _____

7. What kind of ball is there 30 more of than basketballs? _____

Textbook This lesson is available in the eTextbook.

Find the perimeter.

8 A square has one side that is 4 inches long. What is the perimeter?

   a. 16 in.
   b. 12 in.
   c. 8 in.
   d. 4 in.

9 A rectangle has one side that is 8 centimeters long and one side that is 2 centimeters long. What is the perimeter?

   a. 32 cm
   b. 20 cm
   c. 16 cm
   d. 10 cm

10 An equilateral triangle has all sides that are 5 feet long. What is the perimeter?

   a. 20 ft
   b. 15 ft
   c. 10 ft
   d. 5 ft

Ring the measure that is equal.

11 2 ft = _____

   a. 24 in.
   b. 20 in.
   c. 2 yd
   d. 1 yd

12 3 m = _____

   a. 2 yd
   b. 3 yd
   c. 30 cm
   d. 300 cm

13 6 ft = _____

   a. 4 yd
   b. 3 yd
   c. 2 yd
   d. 1 yd

14 500 cm = _____

   a. 1 m
   b. 2 m
   c. 5 m
   d. 50 m

Name _____ Date _____

Use the table to answer the questions.

Children on the Playground	
Age	Number of Children
6	16
7	14
8	20

15 How many children on the playground are 7 years old?

a. 36

b. 20

c. 16

d. 14

16 What is the range of their ages?

a. 8

b. 7

c. 6 to 8

d. 14 to 20

17 What is the mode of their ages?

a. 8

b. 7

c. 6 to 8

d. 14 to 20

Solve for each missing term.

18 5 + _____ = 13

a. 7

b. 8

c. 9

d. 10

19 8 + _____ = 16

a. 6

b. 7

c. 8

d. 9

20 13 − _____ = 10

a. 6

b. 5

c. 4

d. 3

21 17 − _____ = 8

a. 10

b. 9

c. 8

d. 7

Complete the table, and answer the question.

22 Roll a 0–5 **Number Cube** 18 times. Make a tally mark for the number that comes up on each roll. Then write the numbers.

Number	Tallies	Totals
0		
1		
2		
3		
4		
5		

23 **Extended Response** What were the results? Why do you think it came out that way? _____

_____

Find objects that are shorter than 1 meter, about 1 meter long, and longer than 1 meter.

24 **Extended Response** Draw or write to complete the table.

Shorter than 1 Meter	About 1 Meter	Longer than 1 Meter

Thinking Story

TAKE A CHANCE

Decide who has the best chance of being
picked. Write the name.   _____

Skip count by 2s and connect the dots.

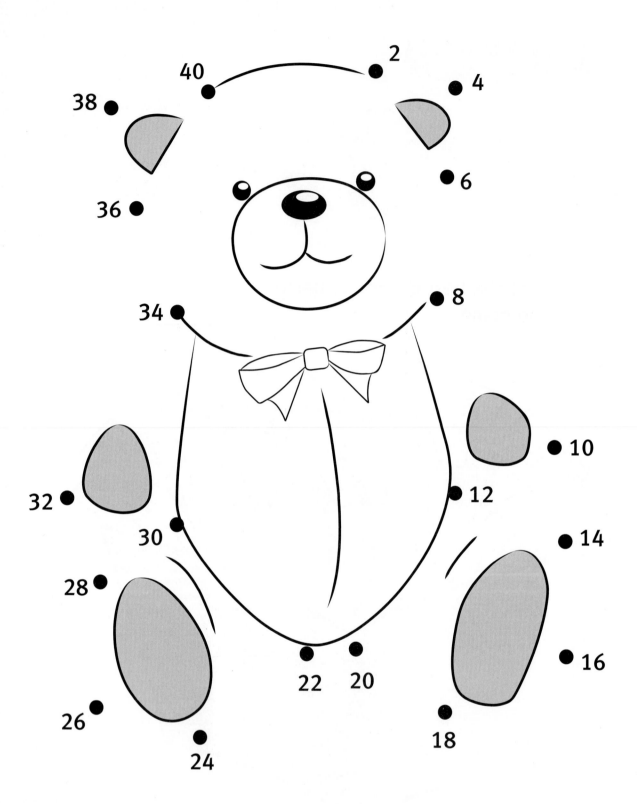

What is this a picture of? _____

Two-Digit Addition

In This Chapter You Will Learn

- to use addition to combine numbers.
- the Commutative Law.
- about function machines.

Problem Solving

Name _____ Date _____

Listen to the problem.

There are ____ spoons.

This is how I organized them.

LESSON 5.1 — Place Value

Key Ideas

A two-digit number tells how many tens and how many ones. 34 means 3 tens and 4 ones.

Each bundle has 10 sticks. There are 3 bundles of 10 and 4 single sticks. There are 34 sticks.

A dime is worth 10¢. A penny is worth 1¢. There are 3 dimes and 4 pennies. There is 34¢.

How many craft sticks? Write your answers.

1  _____

3  _____

2 _____

4 _____

How many cents? Write your answers.

5   _____¢

6   _____¢

e Textbook This lesson is available in the *eTextbook*.

Counting and Strategies Practice

Get to 100 by Tens or Ones Game

Players: Two

Materials: 20 play $10 bills, 20 play $1 bills, paper and pencil for each player

HOW TO PLAY

1 The $10 bills and $1 bills are divided equally between the two players.

2 Beginning with Player One, the players take turns placing one or two bills in the middle of the playing area. If a player runs out of one kind of bill, he or she must use the other kind.

3 Each player keeps a running total of the amount of money put down by counting aloud as each bill is placed. The players should keep written records only if necessary.

4 The player who is first to reach 100 or more wins.

LESSON 5.2 **Adding Tens**

Key Ideas

Adding tens is similar to adding ones.

3 + 5 = 8
3 tens + 5 tens = 8 tens
30 + 50 = 80

Add. Counting by tens will help you do these without using paper and pencil.

1 10 + 0 = _____

2 10 + 10 = _____

3 10 + 10 + 10 = _____

4 20 + 10 = _____

5 10 + 10 + 10 + 10 = _____

6 30 + 10 = _____

7 10 + 10 + 10 + 10 + 10 = _____

8 40 + 10 = _____

9 10 + 10 + 10 + 10 + 10 + 10 = _____

10 50 + 10 = _____

11 10 + 10 + 10 + 10 + 10 + 10 + 10 = _____

12 60 + 10 = _____

e Textbook This lesson is available in the *eTextbook*.

Add. The exercises in one column will help you solve the exercises in the other column.

⓭ 3 + 5 = _____   ⓮ 30 + 50 = _____

⓯ 5 + 2 = _____   ⓰ 50 + 20 = _____

⓱ 8 + 7 = _____   ⓲ 80 + 70 = _____

⓳   8   ⓴   80   ㉑   7   ㉒   70
 + 4   + 40   + 3   + 30

Writing + Math **Journal**

Describe how knowing the sum of 6 + 3 can help you know the sum of 60 + 30.

LESSON 5.3 Regrouping for Addition

Key Ideas

Two bundles of ten and 14 craft sticks can be regrouped as 3 bundles of ten and 4 craft sticks, or 34.

Write the standard name for each of these.

1  _____

5  _____

2  _____

6  _____

3  _____

7  _____

4  _____

8  _____

Write the standard name for each of these.
Use craft sticks if you need help.

9 6 tens and 8 = _____

10 8 tens and 19 = _____

11 1 ten and 8 = _____

12 7 tens and 16 = _____

13 8 tens and 0 = _____

14 0 tens and 8 = _____

15 1 ten and 14 = _____

16 4 tens and 18 = _____

Name _____ Date _____

LESSON 5.4 Adding Two-Digit Numbers

Key Ideas

To add with craft sticks, put bundles of ten together. Put single sticks together. Then make a new bundle if you have enough single sticks.

$48 + 35 =$ _____

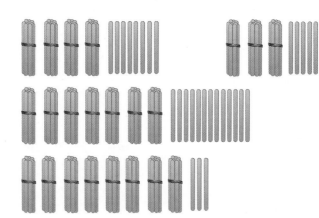

1. two groups

2. combine groups

3. regroup

$48 + 35 = 83$

Use craft sticks or other objects to help add.

1

$50 + 23 =$ _____

There are _____ sticks.

2

$35 + 29 =$ _____

There are _____ sticks.

Use craft sticks or other objects to help add.

③

28 + 24 = ____

There are ____ scoops.

④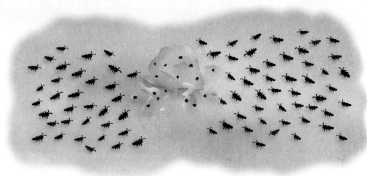

34 + 48 = ____

There are ____ ants.

How many cents?

⑤

34 + 47 = ____   There are ____ cents.

⑥

58 + 35 = ____   There are ____ cents.

Do these exercises. Use any objects you want to help add.

❼ 17 + 18 = ____   ❿ 23 + 15 = ____

❽ 82 + 17 = ____   ⓫ 13 + 68 = ____

❾ 26 + 37 = ____   ⓬ 53 + 42 = ____

LESSON 5.5

More Adding Two-Digit Numbers

Key Ideas

To add two-digit numbers, you can combine tens, combine ones, and then regroup.

$$\begin{array}{r} 31 \\ + 19 \\ \hline \end{array}$$

1. two groups

2. combine groups

3. regroup

$$\begin{array}{r} 31 \\ + 19 \\ \hline 50 \end{array}$$

Use craft sticks or other objects to help add.

1.  46 sticks

 24 sticks

$$\begin{array}{r} 46 \\ + 24 \\ \hline \end{array}$$

Use craft sticks or other objects to help add.

2

27 books

23 books

$$\begin{array}{r} 27 \\ +\ 23 \\ \hline \end{array}$$

3

45¢

25¢

$$\begin{array}{r} 45 \\ +\ 25 \\ \hline \end{array}$$

Add.

4
$$\begin{array}{r} 14 \\ +\ 36 \\ \hline \end{array}$$

5
$$\begin{array}{r} 17 \\ +\ 13 \\ \hline \end{array}$$

6
$$\begin{array}{r} 24 \\ +\ 56 \\ \hline \end{array}$$

7
$$\begin{array}{r} 49 \\ +\ 31 \\ \hline \end{array}$$

8
$$\begin{array}{r} 49 \\ +\ 32 \\ \hline \end{array}$$

9
$$\begin{array}{r} 49 \\ +\ 33 \\ \hline \end{array}$$

10
$$\begin{array}{r} 52 \\ +\ 8 \\ \hline \end{array}$$

11
$$\begin{array}{r} 35 \\ +\ 35 \\ \hline \end{array}$$

12
$$\begin{array}{r} 73 \\ +\ 7 \\ \hline \end{array}$$

Name _____ Date _____

Listen to the problem.

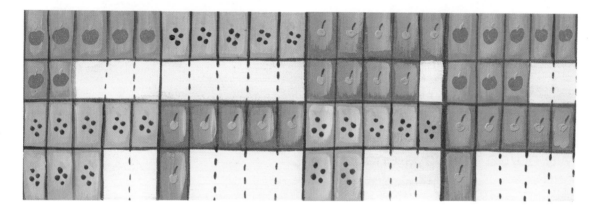

Tanya is acting out the problem.

 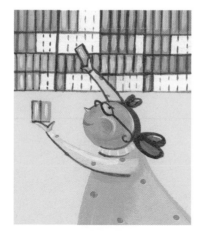 

Exploring Problem Solving

Gilbert is making a diagram to solve the problem.

Ring to show your answer.

We do/do not have enough .

This is how I know.

_____

_____

_____

_____

Cumulative Review

Name _____ Date _____

Counting on a Number Line Lesson 1.7

Solve using the number line. Ring each answer.

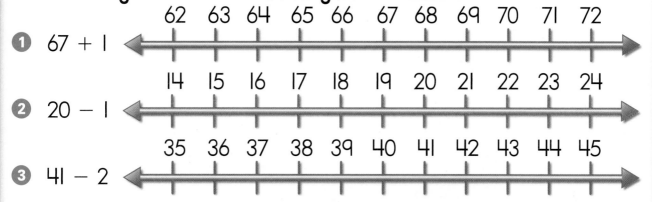

1 67 + 1

2 20 − 1

3 41 − 2

· ·

Collecting and Recording Data Lesson 4.6

Use the table to answer the questions.

How Students Get to School	Number of Students
bus	ⅢⅢ Ⅲ
car	Ⅲ
bike	‖‖‖
walk	Ⅲ ‖

4 How many students walk to school? _____

5 How many students ride the bus? _____

6 How many students get to school on wheels? _____

· ·

Adding Tens Lesson 5.2

Complete the following addition exercises.

7 10 + 10 + 10 + 10 + 10 = _____

8 10 + 10 + 10 + 10 + 10 + 10 = _____

9 10 + 10 + 10 + 10 + 10 + 10 + 10 = _____

10 40 + 10 = _____

11 50 + 10 = _____

12 60 + 10 = _____

Cumulative Review

Horizontal Bar Graphs Lesson 4.11

Use the bar graph to answer the questions.

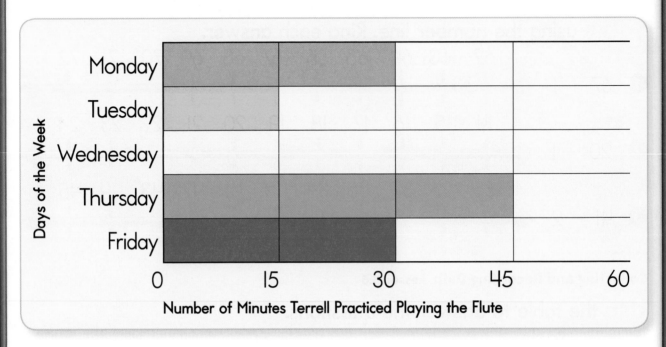

Number of Minutes Terrell Practiced Playing the Flute

⑬ How many minutes did Terrell practice on Tuesday? _____

⑭ How many minutes did Terrell practice on Wednesday? _____

⑮ On which days did Terrell practice for the same number of minutes? _____

⑯ How many minutes did Terrell practice on Saturday? _____

Regrouping for Addition Lesson 5.3

How many sticks?

⑰  _____

⑲  _____

⑱  _____

⑳ _____

LESSON 5.6

Practice with Two-Digit Addition

Key Ideas

When you use craft sticks to add, you can write what you did.

1.   2 tens and 8
   + 3 tens and 5

2.   5 tens and 13

3.   6 tens and 3

Use craft sticks to add. Write what you did.

1   4 tens and 4
   + 3 tens and 6

3   8 tens and 6
   + 0 tens and 7

2   3 tens and 9
   + 1 ten and 0

4   4 tens and 5
   + 5 tens and 3

Add. You may use craft sticks or other objects to help.

5

27 + 16 = _____

There are _____ pretzels.

6

40 + 32 = _____

There are _____ peanuts.

7

35 + 35 = _____

There are _____ cups.

8

30 + 31 = _____

There are _____ days.

LESSON 5.7 More Practice Adding Two-Digit Numbers

Key Ideas

Here is one way to add two two-digit numbers.

$34 + 58 =$ _____

1 3 tens and 4
 + 5 tens and 8

$$\begin{array}{r} 34 \\ + 58 \\ \hline \end{array}$$

2 1 ten
  3 tens and 4
 + 5 tens and 8
  2

$$\begin{array}{r} 1 \\ 34 \\ + 58 \\ \hline 2 \end{array}$$

3 1 ten
  3 tens and 4
 + 5 tens and 8
 9 tens and 2

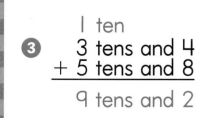

$$\begin{array}{r} 1 \\ 34 \\ + 58 \\ \hline 92 \end{array}$$

Add.

1 89
   + 7

2 19
   + 46

3 15
   + 15

4 25
   + 25

5 45
   + 45

6 94
   + 6

7 0
   + 73

8 30
   + 57

9 89
   + 1

10 37
   + 45

11 38
   + 45

12 39
   + 45

LESSON 5.8 Applications of Two-Digit Addition

Key Ideas

Being able to add two-digit numbers is useful for solving different kinds of problems.

Figure out how much each pair of items costs.

① pen and pencil _____¢

② 2 pens _____¢

③ 2 pencils _____¢

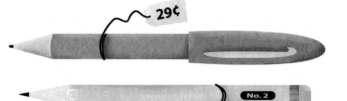

Game Play the **Roll a Problem Game.**

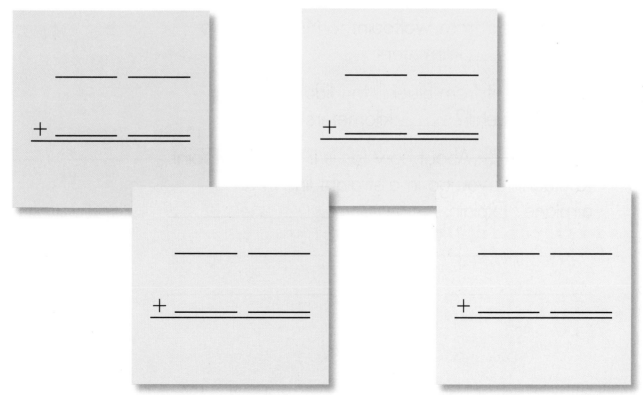

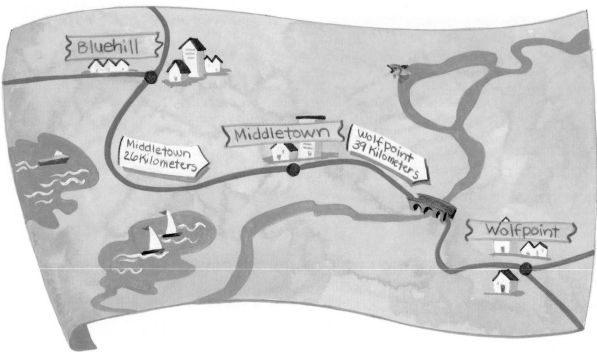

Use the map to answer the questions.

4 How far is it from Bluehill to Middletown to Wolfpoint? _____ kilometers

5 How many people live in Bluehill? _____

6 How far is it from Wolfpoint to Middletown to Bluehill? _____ kilometers

7 How far is it from Bluehill to Middletown and back to Bluehill? _____ kilometers

8 **Extended Response** About how far is it from Wolfpoint to Bluehill if you go in a straight line in an airplane? Explain. _____

_____

_____

Grouping by Tens

Key Ideas

Finding numbers that add to 10 can make a shortcut.

$3 + 4 + 7 + 6 = $ _____

$3 + 7 = 10$ and $4 + 6 = 10$

$10 + 10 = 20$

Add. Use shortcuts if you can.

1 $1 + 3 + 7 + 9 = $ _____

2 $6 + 7 + 4 = $ _____

3 $1 + 3 + 5 + 7 + 9 = $ _____

4 $6 + 7 + 4 + 3 = $ _____

5 $4 + 2 + 6 + 8 + 3 + 7 = $ _____

6 $8 + 2 + 9 + 1 + 7 + 3 + 6 + 4 + 10 = $ _____

7 $4 + 6 + 7 + 3 + 5 + 5 + 9 + 1 = $ _____

8 $1 + 2 + 3 + 4 + 5 + 6 + 7 + 8 + 9 = $ _____

9 $7 + 3 + 1 + 9 + 5 + 5 + 3 = $ _____

10 $5 + 6 + 5 + 4 + 1 + 9 + 7 = $ _____

e Textbook This lesson is available in the *eTextbook*.

Add. Use shortcuts if you can.

⑪ 4 + 1 + 3 + 2 + 5 + 0 + 2 + 3 = _____

⑫ 10 + 9 + 8 + 2 + 1 + 10 = _____

⑬ 2 + 4 + 6 + 8 + 10 = _____

⑭ 20 + 40 + 60 + 80 + 100 = _____

⑮ 21 + 19 + 32 + 18 = _____

⑯ 9 + 1 + 10 + 19 + 1 + 30 = _____

⑰ 42 + 8 + 5 + 6 + 5 + 4 = _____

⑱ Susan worked for five weeks during the summer.
The table shows how much she earned each week.
How much did she earn in those five weeks? Write
the answer in the table.

Week	Dollars Earned
1	41
2	72
3	14
4	28
5	59
Total	

Name _____ Date _____

Addition Shortcuts

Key Ideas

Moving objects from one group to another does not change the number of objects. This idea can be used to find shortcuts for addition.

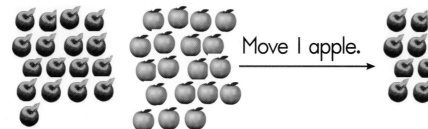 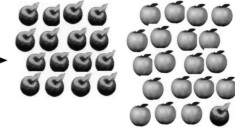

Move 1 apple.

17 + 19 = 36   16 + 20 = 36

Add. Use shortcuts if you can.

1 56 + 29 = _____

2 63 + 19 = _____

3 54 + 39 = _____

4 38 + 42 = _____

5 38 + 47 = _____

6 28 + 32 = _____

Add. Use shortcuts if you can.

7 28 + 37 = _____

10 46 + 39 = _____

8 73 + 19 = _____

11 46 + 38 = _____

9 23 + 19 = _____

12 46 + 37 = _____

13 Extended Response Laura and Anna decided to put their model car collections into one large collection. Anna had 49 cars. Laura had 37 cars. Then they bought 3 more cars together. Now how many cars are in their collection? _____ Explain how you solved the problem. _____

_____

_____

LESSON 5.11

The Paper Explorer—Adding Tens

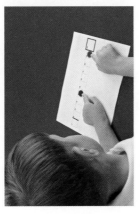

Key Ideas

You can use the Paper Explorer to find sums.

Move both counters at the same time.

Add with the Paper Explorer.

0	10	20	30	40	50	60	70	80	90	100

1. $70 + 50 =$ 120

2. $60 + 50 =$ 110

3. $30 + 40 =$ 70

4. $50 + 90 =$ 140

5. $90 + 90 =$ 180

6. $80 + 70 + 60 =$ 210

REAL WORLD **Solve** these problems.

7 Tanisha had 10¢. Then she earned 60¢ for running an errand. Now how much money does Tanisha have? _____¢

8 Each bottle contains 20 ounces of juice. Ethan bought 3 bottles. How many ounces of juice did he buy? _____ ounces

9 Dexter Park is shaped like a rectangle. One side is 40 meters long, and the other side is 30 meters long. What is the perimeter of Dexter Park? _____ meters

10 Sara is 10 years old. How old will she be in 40 years? _____ years

11 Mark and his mother rode the bus 50 blocks to the library. Then they rode 40 blocks to the supermarket, and then they rode 30 blocks home. How many blocks did they ride the bus? _____

LESSON 5.12 The Paper Explorer—Multidigit Addition

Key Ideas

A counter on one 10 box may be moved to the other without changing the sum. The same is true of the 100 boxes.

1,000	100	10
900	90	9
800	80	8
700	70	7
600	60	6
500	50	5
400	40	4
300	30	3
200	20	2
100	10	1
0	0	0

e Textbook This lesson is available in the *eTextbook*.

Add with the Paper Explorer.
Try to get sums to match those on the page.

1 $3 + 5 = 8$

2 $4 + 8 = 12$

3 $7 + 6 = 13$

4 $8 + 5 = 13$

5 $9 + 3 = 12$

6 $4 + 7 = 11$

7 $30 + 50 = 80$

8 $40 + 80 = 120$

9 $70 + 60 = 130$

10 $80 + 50 = 130$

11 $90 + 30 = 120$

12 $39 + 47 = 86$

13 $92 + 26 = 118$

14 $28 + 31 = 59$

Name _____ Date _____

Listen to the problem.

Mr. García's class has 26 students.

Ms. Taylor's class has 21 students.

Ms. Wilson's class has 25 students.

Each team will have _____ children.

Show how you know.

Ms. Jones's class
has 19 students.

Ms. Jackson's class
has 24 students.

Mr. Clark's class
has 25 students.

Each team will have _____ children.

Show how you know.

Cumulative Review

Name _____ Date _____

Using Multiple Addends Lesson 2.7

Add.

1 $5 + 5 + 4 = $ _____

2 $4 + 8 + 2 = $ _____

3 $10 + 8 + 0 = $ _____

4 $4 + 5 + 9 = $ _____

5 $5 + 6 + 7 = $ _____

6 $1 + 10 + 9 = $ _____

7 $4 + 4 + 4 = $ _____

8 $3 + 4 + 7 = $ _____

Measuring Length—Centimeters Lesson 4.2

Measure the length of each line in centimeters.

9 ▬▬▬▬▬ _____ centimeters

10 ▬▬▬▬▬▬▬ _____ centimeters

Relating Addition and Subtraction Lesson 1.10

Add or subtract.

11 $4 + 6 = $ _____

12 $10 - 6 = $ _____

13 $6 + 4 = $ _____

14 $10 - 4 = $ _____

15 $8 + 5 = $ _____

16 $13 - 5 = $ _____

17 $5 + 8 = $ _____

18 $13 - 8 = $ _____

Cumulative Review

Missing Addends and Subtraction **Lesson 3.2**

Fill in the blanks.

19 7 + _____ = 18

22 7 + _____ = 15

20 14 − _____ = 6

23 14 − _____ = 7

21 9 + _____ = 14

24 17 − 4 = _____

Practice with Two-Digit Addition **Lesson 5.6**

Add.

25 79 + 8 = _____

27 52 + 32 = _____

26 64 + 25 = _____

28 33 + 33 = _____

Adding Tens **Lesson 5.2**

Add.

29 6 + 8 = _____

30 50 + 40 = _____

31 60 + 80 = _____

Addition Shortcuts **Lesson 5.10**

Add. Use shortcuts if you can.

32 20 + 40 + 60 + 80 + 100 = _____

33 6 + 7 + 4 + 3 = _____

34 1 + 3 + 7 + 9 = _____

35 1 + 2 + 3 + 4 + 5 + 6 + 7 + 8 + 9 = _____

Name _____ Date _____

In this chapter you learned about two-digit addition. You learned how to add with craft sticks and how to write what you did. You learned shortcuts for addition.

Answer the questions.

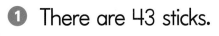

1 There are 43 sticks.

How many tens? _____

How many ones? _____

2 Three tens and 17 is the same as how many? _____

3 You are adding with sticks. You should make a new bundle when you have _____ sticks.

Ring the letter of the correct answer.

Elena is going to add these sticks and write what she did.

4 What should she write first?

   a.   2 tens and 3 tens   **b.**   2 tens and 8   **c.**   5 and 8
      + 8 and 5   + 3 tens and 5   + 3 tens and 2 tens

5 What should she write next?

   a. 13 + 5 tens   **b.** 513 tens   **c.** 5 tens and 13

Answer the questions.

6 Amy and Kwame played the **Roll a Problem Game.**

Amy's problem: 45
                  + 59   Kwame's problem: 45
                                            + 95

Who will win? _____ Why? _____

_____

7 Explain a shortcut for adding $5 + 5 + 4 + 6$.

_____

Ring the letter of the correct answer.

8 $17 + 19$ is the same as

 a. $1 + 7 + 1 + 9$ **b.** $18 + 20$ **c.** $16 + 20$

Use the picture to answer the questions.

9 How many sticks? _____

10 Which of these pictures shows the same number of sticks as in Problem 9?

a.

b.

c.

Name _____ Date _____

Lesson 5.8 **Solve.**

A yo-yo costs 39¢, and a whistle costs 14¢.

1 How much does it cost for both? _____

2 How much do two whistles cost? _____

Lesson 5.2 **Add.**

3 10 + 10 = _____   **6** 30 + 10 = _____

4 20 + 10 = _____   **7** 40 + 20 = _____

5 10 + 10 + 10 = _____ **8** 70 + 90 = _____

Lessons 5.1 and 5.4 **How many cents? Write your answers.**

9 _____ ¢

10 71 + 25 = _____. There are _____¢.

Lessons 5.9–5.10 **Add.** Use shortcuts if you can.

11 $12 + 39 =$ _____

15 $29 + 27 =$ _____

12 $41 + 29 =$ _____

16 $35 + 45 =$ _____

13 $23 + 46 =$ _____

17 $52 + 16 =$ _____

14 $16 + 14 =$ _____

18 $36 + 39 =$ _____

19 $1 + 3 + 5 + 7 + 9 =$ _____

20 $9 + 1 + 10 + 19 + 1 + 30 =$ _____

21 The pool in the park is shaped like a rectangle. One side is 50 meters long, and another side is 25 meters long. What is the perimeter of the pool?

_____ meters

Lesson 5.6 **Add.** Write what you did.

22   3 tens and 7
   $+$ 5 tens and 2
   _____
      or _____

23   5 tens and 1
   $+$ 3 tens and 9
   _____
      or _____

Practice Test

Name _____ Date _____

How many craft sticks? Write your answers.

1

_____

2

_____

3

_____

Write the standard name for each of these.

4

_____

5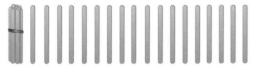

_____

6

_____

Add.

7 $\begin{array}{r} 7 \\ +\ 23 \\ \hline \end{array}$ **8** $\begin{array}{r} 15 \\ +\ 55 \\ \hline \end{array}$ **9** $\begin{array}{r} 14 \\ +\ 76 \\ \hline \end{array}$ **10** $\begin{array}{r} 21 \\ +\ 39 \\ \hline \end{array}$

e Textbook This lesson is available in the *eTextbook*.

Practice Test

Add. Count by tens.

⑪ 30 + 40 = _____

   a. 7   **b.** 10

   c. 60   **d.** 70

⑬ 50 + 20 = _____

   a. 70   **b.** 60

   c. 40   **d.** 30

⑫ 80 + 10 = _____

   a. 90   **b.** 80

   c. 70   **d.** 60

⑭ 60 + 20 = _____

   a. 90   **b.** 80

   c. 50   **d.** 40

Find the measure that is equal.

⑮ 36 in. = _____

   a. 3 yd   **b.** 2 ft

   c. 1 ft   **d.** 1 yd

⑰ 7 m = _____

   a. 700 cm **b.** 70 cm

   c. 700 ft   **d.** 70 ft

⑯ 2 ft = _____

   a. 30 in.   **b.** 24 in.

   c. 20 in.   **d.** 12 in.

⑱ 200 cm = _____

   a. 1 m   **b.** 2 m

   c. 20 m   **d.** 200 m

Practice Test

Name _____ Date _____

Add.

19 33 + 37 = _____

    a. 60   **b.** 64   **c.** 70   **d.** 74

20 72 + 28 = _____

    a. 100   **b.** 80   **c.** 56   **d.** 46

21 16 + 15 = _____

    a. 20   **b.** 21   **c.** 30   **d.** 31

Solve.

22 Yoshi drove 30 miles to visit her aunt. Then she drove 20 miles to visit her grandfather. The next day, she drove 40 miles to get home. How many miles in all did she drive?

    a. 90   **b.** 70   **c.** 60   **d.** 50

23 Helen bought 2 shirts for $20 each and 1 dress for $30. How much did Helen spend?

    a. $40   **b.** $50   **c.** $70   **d.** $80

24 Nick spent 34¢ on a pad of paper and 27¢ on a pen. How much did he spend in all?

    a. 51¢   **b.** 61¢   **c.** 63¢   **d.** 67¢

e Textbook This lesson is available in the *eTextbook*.

Use the map to solve.

25. Find three routes to get from Afton to Brownsville.
Find the total miles for each route. Show your work.

_____

_____

_____

Add. Describe the shortcuts you used.

26. 23 + 35 + 17 + 5 + 0 + 2 + 6 + 2

_____

_____

_____

Thinking Story

Sharing
with
Cousin Trixie

Put an X on the extra grapes from napkins
that have too many. Draw grapes on napkins
that have too few. Write the number of
grapes that should be on each napkin if they
are shared evenly.

_____

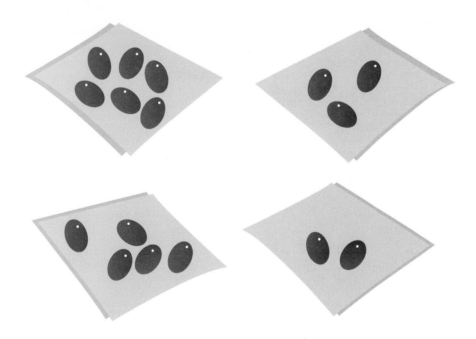

You and three friends have an hour to share a bicycle. Color the amount of time each person would have to ride.

Two-Digit Subtraction

In This Chapter You Will Learn

- regrouping.
- subtracting from tens.
- two-digit subtraction.

Name _____ Date _____

Listen to the problem. Draw your design.

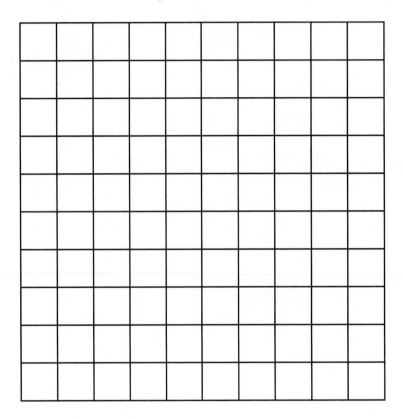

How much space will the pond take?   _____ squares

How big will the area for the workers be?   _____ squares

How much space will be left for the polar
bears to walk?   _____ squares

How do you know? _____

_____

_____

Real Math • Chapter 6

Name _____ Date _____

Regrouping Tens as Ones

Key Ideas

You can regroup craft sticks to show one fewer ten and ten more ones.

55 = 4 tens and 15

Rewrite each number to show one fewer ten.

① 63 = ____ tens and ____

② 49 = ____ tens and ____

③ 38 = ____ tens and ____

④ 43 = ____ tens and ____

⑤ 60 = ____ tens and ____

⑥ 22 = ____ ten and ____

⑦ 36 = ____ tens and ____

⑧ 15 = ____ tens and ____

⑨ 78 = ____ tens and ____

⑩ 46 = ____ tens and ____

Rewrite each number to show one fewer ten.

⑪ 50 = _____ tens and _____

⑫ 48 = _____ tens and _____

⑬ 100 = _____ tens and _____

⑭ 89 = _____ tens and _____

⑮ 56 = _____ tens and _____

⑯ 97 = _____ tens and _____

⑰ 61 = _____ tens and _____

⑱ 14 = _____ tens and _____

⑲ 37 = _____ tens and _____

⑳ 28 = _____ ten and _____

 **Writing + Math** **Journal**

Explain how to rewrite the number 73 to show one fewer ten.

LESSON 6.2 Subtracting Tens

Key Ideas

Subtracting groups of tens is like subtracting ones.

$60 - 20 = 40$ 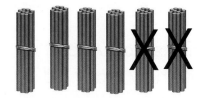 $6 - 2 = 4$

Solve these problems.

1. Mika had $30. She spent $20 at the zoo.
   Now she has $____.

2. Jerry has $50. The skateboard costs $80.
   How much more money does he need
   to buy the skateboard? He needs $____.

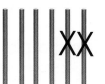

3. $30 - 10 =$ ____

4. $80 - 20 =$ ____

Subtract or add.

5.
$$\begin{array}{r} 60 \\ + 30 \\ \hline \end{array}$$

6.
$$\begin{array}{r} 90 \\ - 40 \\ \hline \end{array}$$

7.
$$\begin{array}{r} 90 \\ - 50 \\ \hline \end{array}$$

8.
$$\begin{array}{r} 80 \\ - 50 \\ \hline \end{array}$$

9.
$$\begin{array}{r} 80 \\ - 60 \\ \hline \end{array}$$

10.
$$\begin{array}{r} 70 \\ - 60 \\ \hline \end{array}$$

Subtract or add.

⑪ 30 + 40 =   ⑬ 70 − 20 =

⑫ 70 − 30 =   ⑭ 50 + 20 =

Solve these problems.

⑮ Chandra planted about 70 radish seeds. About 50 of the seeds sprouted. About how many did not sprout? _____

⑯ Sara's goal is to collect 100 different baseball cards. So far she has collected 40 of them. How many more cards must she collect to reach her goal? _____

⑰ Andrea needed $40 to buy a bicycle. She earned $10 babysitting. Then she earned some more money weeding gardens. Does she have enough money now?

_____

⑱ Heide bought a basketball for $30. She gave the clerk 2 $20 bills. How much change should she get? _____

⑲ A pine tree was about 80 feet tall. A storm came, and the top 10 feet broke off. About how tall is the tree now? _____

⑳ Mr. Joseph planned to drive about 100 miles from San Diego to Long Beach. So far, he has driven about 50 miles. About how much farther must he drive?

_____

LESSON 6.3

Subtracting from Tens

Key Ideas

You can use regrouping to help you subtract. $60 - 37 = ?$

$$\begin{array}{r} 60 \\ -37 \\ \hline \end{array} = \begin{array}{r} 6 \text{ tens and } 0 \\ -3 \text{ tens and } 7 \\ \hline \end{array} = \begin{array}{r} 5 \text{ tens and } 10 \\ -3 \text{ tens and } 7 \\ \hline 2 \text{ tens and } 3, \text{ or } 23 \end{array}$$

Subtract. Use craft sticks if you need help. Record what you did.

1 $\begin{array}{r} 90 \\ -48 \\ \hline \end{array}$

2 $\begin{array}{r} 80 \\ -30 \\ \hline \end{array}$

3 $\begin{array}{r} 70 \\ -48 \\ \hline \end{array}$

4 $\begin{array}{r} 60 \\ -17 \\ \hline \end{array}$

5 $\begin{array}{r} 50 \\ -25 \\ \hline \end{array}$

6 $\begin{array}{r} 40 \\ -10 \\ \hline \end{array}$

7 70 − 29 =   **8** 50 − 38 =   **9** 90 − 70 =

10 80 − 40 =   **11** 60 − 44 =   **12** 50 − 20 =

13   40   **14**   80   **15**   100
     − 12   − 52   − 75

16 Extended Response How did you find the answer to
Exercise 15? _____

17 Ravi bought a DVD player that had a regular price
of $60. It was on sale for $42. How much did Ravi
save? _____

18 Extended Response The Kelly twins bought two of the DVD
players. How much did they save together? _____
How did you find the answer? _____

 **Journal**

Why do you sometimes need to regroup when
subtracting a two-digit number from a group of tens?

LESSON 6.4 Subtracting Two-Digit Numbers

Key Ideas

Use what you learned about subtracting from tens to subtract from any two-digit number.

$$53 - 25 = ?$$

$$\begin{array}{r} 5\,3 \\ -\ 2\,5 \\ \hline \end{array}$$ ➡
5 tens and 3
— 2 tens and 5

$$\begin{array}{r} {}^{4}\!\!\not{5}\,13 \\ -\ 2\,5 \\ \hline \end{array}$$ ➡
4 tens and 13
— 2 tens and 5
2 tens and 8, or 28

Subtract.

1 62 − 14 =

3 62 − 16 =

2 62 − 15 =

4 62 − 17 =

5
```
   83
 − 73
```

9
```
   33
 − 17
```

6
```
  100
 − 25
```

10
```
   63
 − 37
```

7
```
   17
 − 17
```

11
```
   18
 −  9
```

8
```
   92
 − 49
```

12
```
   26
 − 13
```

212

LESSON 6.5 Using a Paper Explorer to Understand Subtraction

Key Ideas

The Paper Explorer can be used to solve subtraction problems.

Use the Paper Explorer your teacher gives you. Work together to solve these problems.

1,000	100	10
900	90	9
800	80	8
700	70	7
600	60	6
500	50	5
400	40	4
300	30	3
200	20	2
100	10	1
0	0	0

1 50 − 27 = 23

2 80 − 60 = 20

3 99 − 44 = 55

4 57 − 23 = 34

5 48 − 39 = 9

6 40 − 27 = 13

7 74 − 13 = 61

8 100 − 93 = 7

9 55 − 10 = 45

10 60 − 15 = 45

ⓔTextbook This lesson is available in the *eTextbook*.

Solve these problems. Explain your answers.

Marta has $50. Enrique has $32.

11 `Extended Response` Can Enrique buy the bicycle? Explain. _____

_____

12 `Extended Response` Could Marta and Enrique buy the bicycle together? Explain. _____

_____

13 If Marta and Enrique buy the bicycle together, how much money will they have left?

_____

14 If Marta buys the kite, how much money will she have left? _____

15 `Extended Response` If Enrique buys a baseball and bat, how much money will he have left? _____

_____

Name _____ Date _____

Practicing Two-Digit Subtraction

Key Ideas

The same steps work for subtracting any numbers.

Remember to look at the ones place before subtracting.

$$
\begin{array}{r} 83 \\ -25 \\ \hline \end{array}
\quad\Longrightarrow\quad
\begin{array}{r} \overset{7}{\cancel{8}}13 \\ -25 \\ \hline \end{array}
\quad\Longrightarrow\quad
\begin{array}{r} \overset{7}{\cancel{8}}13 \\ -25 \\ \hline 58 \end{array}
$$

Subtract. Write your answers.

1. $\begin{array}{r} 70 \\ -43 \\ \hline \end{array}$
2. $\begin{array}{r} 73 \\ -43 \\ \hline \end{array}$
3. $\begin{array}{r} 74 \\ -43 \\ \hline \end{array}$
4. $\begin{array}{r} 68 \\ -39 \\ \hline \end{array}$
5. $\begin{array}{r} 81 \\ -26 \\ \hline \end{array}$

Solve these problems.
Use play money if you need to.

6. David had $63. He spent $28. Now he has _____.

7. Yori has $43. She needs $62. She needs _____ more.

8. **Extended Response** Jacob has $43. Emily has $39. Do they have enough money to buy five sports tickets? _____

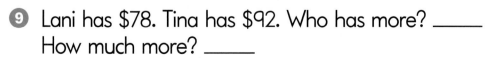

9. Lani has $78. Tina has $92. Who has more? _____ How much more? _____

10. Li had $27. She earned $18. Now she has _____.

e Textbook This lesson is available in the *eTextbook*.

Place Value, Subtraction, and Strategies Practice

Roll a Problem (Subtraction) Game

Players: Two or more

Materials: Paper and pencil for each player, *Number Cube* (0–5)

HOW TO PLAY

1 Begin by drawing lines to represent a two-digit subtraction problem on a piece of paper, like this:

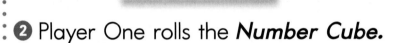

2 Player One rolls the *Number Cube.*

3 Player One announces the digit rolled, and all players write that digit in one of the spaces before the next number is rolled.

4 After players have a completed problem, they subtract to find the difference.

5 In each round, the player who has the number closest to, but not less than, zero is the winner. There may be ties.

Name _____ Date _____

Listen to the problem.

Ms. Smith needs 60 fish to feed the sea lions.

How many more fish does she need?

Owen made a plan to solve the problem.

1. Find how many fish are in the buckets.
2. If there are fewer than 60,
   find how many more to make 60.

Harper is drawing a picture to solve the problem.

Fish she needs

| 10 | 10 | 10 | 10 | 10 | 10 |

Fish she has

Show how you solved the problem.

Ms. Smith needs _____ more fish.

Cumulative Review

Name _____ Date _____

Odds and Evens Lesson 1.2

Ring the odd numbers. Write how many would
be in each half if you split the even numbers.

1 64   **2** 13   **3** 25   **4** 46

5 47   **6** 36   **7** 16   **8** 24

Adding Two-Digit Numbers Lesson 5.4

Add.

9   34
+ 17

10   14
+ 28

11   16
+ 34

12   65
+ 15

13   52
+ 35

14

35
+ 23

There are _____ penguins.

Regrouping Tens as Ones Lesson 6.1

Rewrite each number to show one fewer ten and ten more ones.

15 40 = _____ tens and _____   **18** 50 = _____ tens and _____

16 20 = _____ ten and _____   **19** 48 = _____ tens and _____

17 37 = _____ tens and _____   **20** 56 = _____ tens and _____

Cumulative Review

Subtracting Two-Digit Numbers Lesson 6.4

Solve these problems.

21 40
− 30

22 30
− 20

23 80
− 30

24 40
− 28

25 70
− 28

26 90
− 71

27 80
− 20

28 51
− 15

Addition and Subtraction Functions Lesson 3.3

Fill in the missing numbers.

29 in −11 out

19	8
43	32
	18
58	

30 in −25 out

36	11
	20
54	
34	9

31 in out

49	57
27	
16	24
	23

The rule is _____

Grouping by Tens Lesson 5.9

Add. Use shortcuts if you can.

32 10 + 4 + 4 = _____

35 1 + 3 + 7 + 9 = _____

33 6 + 4 + 8 + 1 = _____

36 21 + 19 + 32 + 18 = _____

34 6 + 7 + 4 + 3 = _____

37 6 + 7 + 4 = _____

LESSON 6.7 Checking Subtraction

Key Ideas

You can check a subtraction answer by using addition.

$16 - 4 = 12$

Check your solution by adding 12 to 4.
The correct answer is 12.

$12 + 4 = 16$

Check using addition. Ring each wrong answer. Write the correct answers.

1
$$\begin{array}{r} 17 \\ -\ 8 \\ \hline 9 \end{array}$$

2
$$\begin{array}{r} 84 \\ -\ 27 \\ \hline 57 \end{array}$$

3
$$\begin{array}{r} 61 \\ -\ 34 \\ \hline 26 \end{array}$$

4
$$\begin{array}{r} 70 \\ -\ 43 \\ \hline 27 \end{array}$$

5
$$\begin{array}{r} 29 \\ -\ 23 \\ \hline 52 \end{array}$$

6
$$\begin{array}{r} 40 \\ -\ 32 \\ \hline 18 \end{array}$$

7
$$\begin{array}{r} 100 \\ -\ 70 \\ \hline 30 \end{array}$$

8
$$\begin{array}{r} 45 \\ -\ 25 \\ \hline 18 \end{array}$$

9
$$\begin{array}{r} 64 \\ -\ 35 \\ \hline 38 \end{array}$$

10
$$\begin{array}{r} 75 \\ -\ 25 \\ \hline 50 \end{array}$$

11
$$\begin{array}{r} 75 \\ -\ 26 \\ \hline 49 \end{array}$$

12
$$\begin{array}{r} 75 \\ -\ 27 \\ \hline 38 \end{array}$$

13
$$\begin{array}{r} 87 \\ -\ 38 \\ \hline 49 \end{array}$$

14
$$\begin{array}{r} 76 \\ -\ 35 \\ \hline 35 \end{array}$$

15
$$\begin{array}{r} 92 \\ -\ 13 \\ \hline 79 \end{array}$$

16
$$\begin{array}{r} 70 \\ -\ 43 \\ \hline 27 \end{array}$$

Solve these problems.

17 Arnold's digital camera can hold  85 pictures. He has already taken 46 pictures. How many more can he take on his trip to the zoo? _____

18 Toothpaste costs 84¢.
The toothbrush costs 79¢.
Which costs more? _____
How much more? _____

19 Rosa has $95. Janet has $59.
Who has more? _____
How much more? _____

20 Antonio had 73 football cards.
He gave some away. Now he has 45 cards.
How many did he give away? _____

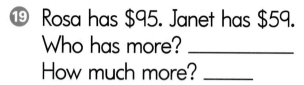 **Journal**

Write your own word problem that uses two-digit subtraction.

LESSON 6.8 **Applying Subtraction**

Key Ideas

Subtraction can help you solve problems every day. How much money will she have left if she buys the shirt?

$40 - 27 = 13$

Find the answers.

1
$$\begin{array}{r} 50 \\ -\ 25 \\ \hline \end{array}$$

2
$$\begin{array}{r} 50 \\ -\ 26 \\ \hline \end{array}$$

3
$$\begin{array}{r} 50 \\ -\ 27 \\ \hline \end{array}$$

4
$$\begin{array}{r} 50 \\ -\ 28 \\ \hline \end{array}$$

5 CDs are on sale at three for $2. How much will 9 CDs cost? _____ Explain. _____

_____

6 Jeremy is allowed to use his cell phone for 65 minutes each week. He has already used 49 minutes. Does Jeremy have enough time to make a 15-minute phone call? _____

Solve these problems using the picture. Explain your answers.

7 Nathan had $76. He bought a pair of shoes. He now has _____.
Explain. _____

8 Rishi bought a shirt. He gave the clerk two $20 bills. How much change did he get? _____

9 Extended Response Simon has $90. He wants to buy a jacket and pants. Can he? _____
Explain. _____

10 Stan has a $50 bill. He buys pants and a belt. How much change should he get? _____

11 Tyrone bought a shirt, pants, and a belt. He gave the clerk four $20 bills. How much change should he get? _____

12 Look at the storefront picture.
Make up your own problem. Solve it.

_____

_____

_____

 **Journal**

Can you list some ways that you or your family uses subtraction?

LESSON 6.9 Comparing Two-Digit Subtraction Expressions

Key Ideas

> is greater than, < is less than, = is equal to

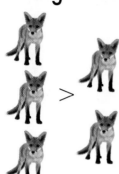 >

 <

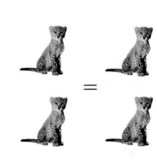 =

What is the correct sign? Draw >, <, or =.

1 27 ◯ 14

2 35 ◯ 62

3 97 ◯ 97

4 18 ◯ 46

Write a number to make the following sentences correct.

5 35 > _____ − 42

6 32 + _____ > 57

7 17 > 29 − _____

8 98 − _____ < 49

What is the correct sign? Draw >, <, or =.

9 42 − 14 ◯ 28

10 56 ◯ 70 − 14

11 15 + 15 ◯ 28

12 54 ◯ 54 + 30

13 36 − 8 ◯ 28

14 64 ◯ 68 − 7

15 4 + 12 ◯ 28

16 54 ◯ 17 + 37

ⓔTextbook This lesson is available in the *eTextbook*.

The Peláez family drove 37 miles from home to the national park. Then they drove 25 miles to the museum.

17 **Extended Response** How far is it from their home to the museum?

_____

Explain. _____

_____

18 How far is it from the national park to the museum? _____

Explain. _____

_____

19 How far did the Peláez family drive before they got to the museum? _____

Explain. _____

Writing + Math **Journal**

Write each relation sign, and give an example of when it is used to show the relationship between two numbers.

Name _____   Date _____

Key Ideas

Mathematics can be used to explain situations in our world. The boy can use the map to tell how far it is from the reptile house to the aviary.

Extended Response **Look** at the map. Answer these questions.

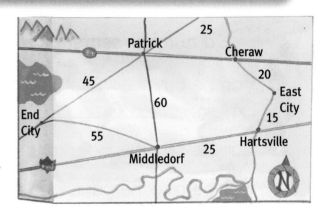

1 Mr. Jones wants to drive from Cheraw to Middledorf. What is the shortest driving distance? _____ Explain. _____

2 Mr. Patel wants to drive from East City to End City. What is the shortest driving distance? _____ Explain. _____

3 Ms. Li lives in Patrick and works in Middledorf. She drives back and forth once each day. How many miles is that? _____ Explain. _____
_____

4 If a bird were to fly from East City to End City, about how many miles would it fly?
About _____.

Use information from the map above.
Create your own word problem.

5 _____

Textbook This lesson is available in the *eTextbook*.

Mr. Meier has $90.

Write a number sentence for each question. Solve.

6 Can he buy the jacket and pants?
Explain. _____

7 Can he buy the jacket and hat? _____

8 Can he buy the pants and hat? _____
How much money will he have left after buying the
pants and the hat? _____

Fill in the missing term.

9 $25 + \underline{\hspace{1cm}} = 88$

12 $\underline{\hspace{1cm}} - 43 = 44$

10 $\underline{\hspace{1cm}} - 39 = 42$

13 $41 - \underline{\hspace{1cm}} = 18$

11 $64 + \underline{\hspace{1cm}} = 91$

14 $\underline{\hspace{1cm}} + 17 = 47$

Exploring Problem Solving

Name _____ Date _____

Listen to the problem.

Plan your morning visit. Use the table to help you.

Habitat	Walking Time	Looking Time	Total Time	Time Left

I will get back at _____.

Plan your afternoon visit. Use this table to help you.

Habitat	Walking Time	Looking Time	Total Time	Time Left

I will get back at _____.

Cumulative Review

Name _____ Date _____

Checking Subtraction Lesson 6.7

Check using addition. **Ring** each wrong answer.
Write the correct answers alongside the problem.

1 50 − 20 = 30

2 90 − 25 = 50

3 47 − 21 = 26

4 35 − 13 = 48

5 55 − 31 = 86

6 33 + 25 = 58

The Calendar Lesson 1.6

Fill in the missing numbers. Then answer the questions.
October has 31 days.

SUN	MON	TUE	WED	THU	FRI	SAT
1	2	3	4	5		
		17				

October

7 What day is October 15? _____

8 What date is the fourth Monday? _____

9 What day is October 31? _____

10 What day is the second Saturday? _____

Cumulative Review

Regrouping Tens as Ones Lesson 6.1

Rewrite each number to show one fewer ten and ten more ones.

⑪ 68 = _____ tens and _____

⑭ 14 = _____ tens and _____

⑫ 76 = _____ tens and _____

⑮ 37 = _____ tens and _____

⑬ 61 = _____ tens and _____

⑯ 28 = _____ ten and _____

Chain Calculations Lesson 3.7

Write the answers.

It takes Adam about 6 minutes to feed 2 elephants.

⑰ About how long would it take Adam to feed 6 elephants? About _____ minutes

⑱ About how long would it take Adam to feed 9 elephants? About _____ minutes
Explain. _____

_____

Name _____ Date _____

In this chapter you learned about two-digit subtraction. You used subtraction to solve everyday problems. You learned how to use addition to check your solution to a subtraction problem.

· ·

Ring the letter of the correct answer.

① How can you rewrite 55?

  a. 4 tens and 5   **b.** 4 tens and 15

  c. 6 tens and 15

② Which picture below does not show 36 sticks?

  a.   **b.**   **c.**

Write your answers.

Shane is going to use craft sticks to solve 60 − 37. He will write what he did.

③ What should he write first?

  _____ tens and _____

  − _____ tens and _____

④ What should he write now?

  _____ tens and _____

  − _____ tens and _____

⑤ What answer will Shane get? _____

Key Ideas Review

Ring the letter of the number sentence you would use to solve each problem.

6 David had $63. He spent $28. How much money does he have now?

   a. $63 - 28 =$

   b. $28 - 63 =$

   c. $63 + 28 =$

7 Cruz has 2 dimes and 7 pennies. A pencil costs 35¢. How much more money does Cruz need to buy a pencil?

   a. $27 - 35 =$

   b. $35 - 27 =$

   c. $27 + 35 =$

Write your answers.

8 Janet and Erin played the **Roll a Problem Game.**

Janet's problem looked like this:
$$\begin{array}{r} 51 \\ -\ 24 \\ \hline \end{array}$$

Erin's problem looked like this:
$$\begin{array}{r} 45 \\ -\ 21 \\ \hline \end{array}$$

   a. Solve each problem.

   b. Who had the difference closer to 0? _____

9 Andre subtracted 14 from 37 and got 23. How can he check his answer? _____

_____

10 Myra picked 47 apples. She gave some away. Now she has 19 apples. How many did she give away? _____

Name _____ Date _____

Lessons 6.6–6.7 **Check** using addition. Ring each wrong answer. Then write the correct answers alongside the problem.

1 $48 - 2 = 46$

2 $71 - 24 = 95$

3 $31 - 24 = 6$

4 $51 - 43 = 18$

5 $80 - 22 = 58$

6 $39 - 33 = 62$

Solve this problem.

7 Adam has $87. Matt has $59.

Who has more? _____ How much more? _____

Lesson 6.1 **Rewrite** each number to show one fewer ten and ten more ones.

8 $54 = $ ___ tens and ___

9 $33 = $ ___ tens and ___

10 $100 = $ ___ tens and ___

11 $49 = $ ___ tens and ___

12 $27 = $ ___ ten and ___

13 $89 = $ ___ tens and ___

Lesson 6.9 **What** is the correct sign? Draw >, <, or =.

14 $42 - 14 \bigcirc 28$

15 $64 \bigcirc 68 - 9$

Lessons 6.2–6.4 **Subtract.** Use craft sticks or play money to help solve these problems. Record what you did.

⑯ Sophie had $70. She spent $20 at the fair. Now she has $_____.

⑰ Tim has $68. Sheila has $96. Who has more? _____ How much more? _____

⑱
```
   50
 − 30
```

⑲
```
   80
 − 20
```

⑳
```
   90
 − 50
```

㉑
```
   80
 − 68
```

㉒
```
   40
 − 17
```

㉓
```
   42
 − 12
```

Lessons 6.8 and 6.10 **Write** a number sentence for each question.

PLANT FOOD $8

$54 $26

Mr. Franklin has $85.

㉔ Can Mr. Franklin buy the hose and the nozzle? Explain.

_____

_____

㉕ Can he buy all three items? _____

_____

Name _____ Date _____

Rewrite each number to show one fewer ten.

1 40 = _____ tens and _____

2 78 = _____ tens and _____

3 31 = _____ tens and _____

4 55 = _____ tens and _____

Subtract.

5
$$\begin{array}{r} 30 \\ -\ 20 \\ \hline \end{array}$$

6
$$\begin{array}{r} 50 \\ -\ 10 \\ \hline \end{array}$$

7
$$\begin{array}{r} 80 \\ -\ 30 \\ \hline \end{array}$$

8
$$\begin{array}{r} 60 \\ -\ 40 \\ \hline \end{array}$$

Subtract. Then use addition to check.

9
$$\begin{array}{r} 86 \\ -\ 28 \\ \hline \end{array}$$
$+\ \square$

10
$$\begin{array}{r} 51 \\ -\ 33 \\ \hline \end{array}$$
$+\ \square$

Practice Test

Subtract.

11 70 − 14 = _____
  a. 66
  b. 56
  c. 54
  d. 44

13 87 − 39 = _____
  a. 52
  b. 48
  c. 42
  d. 38

12 40 − 22 = _____
  a. 28
  b. 22
  c. 18
  d. 12

14 66 − 47 = _____
  a. 19
  b. 24
  c. 29
  d. 34

Ring a number to make each sentence true.

15 26 + 46 > _____
  a. 71
  b. 72
  c. 73
  d. 74

17 54 − 28 > _____
  a. 28
  b. 27
  c. 26
  d. 25

16 72 − 6 < _____
  a. 64
  b. 65
  c. 66
  d. 67

18 43 + 17 = _____
  a. 74
  b. 70
  c. 60
  d. 54

Ferdie Borrows and Borrows and Borrows

Ferdie has 40¢. Write how much money he would have to borrow to buy each toy shown.

60¢

$1

_____

40¢

_____

80¢

_____

Draw a line from the money Ferdie owes to the job he could do to earn the right amount of money to pay it back.

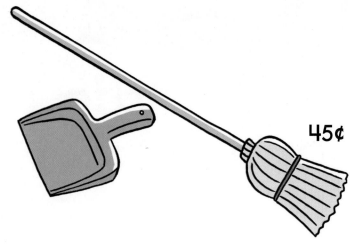

45¢

50¢

8¢

In This Chapter You Will Learn

- what a fraction is.
- how to create and use fractions.

Problem Solving

Name _____ Date _____

Listen to the problem.

Show how you made three equal portions.

```
┌─────────────────────────────────────────┐
│                                         │
│                                         │
│                                         │
│                                         │
│                                         │
│                                         │
│                                         │
└─────────────────────────────────────────┘
```

How do you know the three portions are equal?

_____

_____

_____

Real Math • Chapter 7

LESSON 7.1

Halves and Fourths

Key Ideas

When something has two equal parts, each part is <mark>one-half</mark> of the whole.

When something has four equal parts, each part is <mark>one-fourth</mark> of the whole.

Listen and follow the directions.

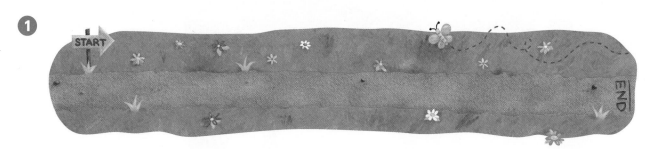

1

2

3

4 Draw lines to show one-half of each ribbon.

5 Extended Response The beginnings of all the ribbons are lined up. Are the halfway marks lined up? _____ Why or why not? _____

_____

6 Draw a line on each glass to show what it would look like if it were half full.

7 Extended Response Do the two half glasses of juice hold the same amount? _____
Why or why not? _____

_____

LESSON 7.2 Writing Fractions

Key Ideas

When you write a fraction, you write a number above the line and a number below the line.

The number below the line, or **denominator**, tells how many equal parts there are. The number above the line, or **numerator**, tells how many of those parts you are talking about.

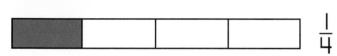 $\frac{1}{2}$

$\frac{1}{4}$

What fraction of each bar is shaded?

1  _____

2 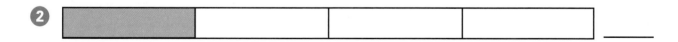 _____

3 This shaded part shows $\frac{1}{2}$ of a rectangle.
Draw the rest of the rectangle.

Jai and Liz are riding their bikes to the museum.
The path to the museum is 8 kilometers long.

④ Jai has gone $\frac{1}{2}$ of the way.
How many kilometers has he gone? _____

⑤ Liz has gone $\frac{1}{4}$ of the way.
How many kilometers has she gone? _____

⑥ Draw a *J* to show where Jai is.
Draw an *L* to show where Liz is.

⑦ How many more kilometers does Jai
have to ride to reach the museum? _____

⑧ How many more kilometers does Liz
have to ride to reach the museum? _____

 **Journal**

Why is $\frac{1}{2}$ of something more than $\frac{1}{4}$ of the same thing?

LESSON 7.3

Halves, Fourths, and Thirds

Key Ideas

Halves	Fourths	Thirds

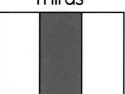

$\frac{1}{2}$ is shaded green. $\frac{1}{4}$ is shaded red. $\frac{1}{3}$ is shaded blue.

1 Color $\frac{1}{2}$ of each figure.

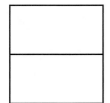

4 Color $\frac{1}{3}$ of each figure.

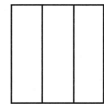

2 Color $\frac{1}{4}$ of each figure.

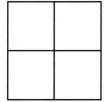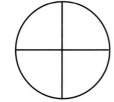

5 Color $\frac{2}{3}$ of each figure.

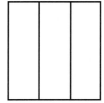

3 Color $\frac{2}{4}$ of each figure.

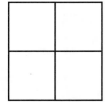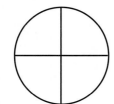

6 Color $\frac{3}{4}$ of each figure.

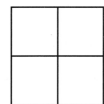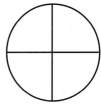

e Textbook This lesson is available in the *eTextbook*.

Laura's class studied flags of the world.
This is the flag of Ireland:

$\frac{1}{3}$ of the flag is green. $\frac{1}{3}$ of the flag is white.

$\frac{1}{3}$ of the flag is orange.

Students in Laura's class designed their own flags.
Color the rectangles to show what each flag might
look like.

7 Jeff colored $\frac{1}{2}$ of his flag purple.

8 Consuela drew a symbol
on $\frac{1}{4}$ of her flag.

9 Emmy made a flag that
was $\frac{3}{4}$ blue and $\frac{1}{4}$ green.

LESSON 7.4 Sixths and Eighths

Key Ideas

Sixths

Eighths

Each string is a fraction of the length of the red string.
Match each string with the correct fraction.

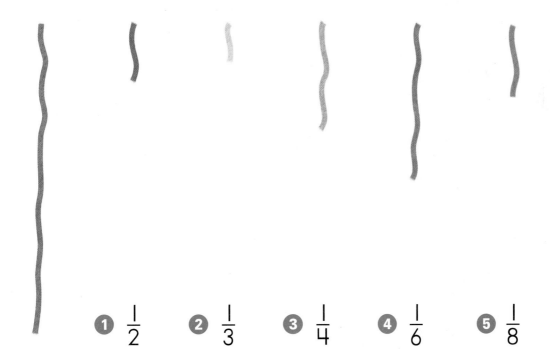

1 $\dfrac{1}{2}$ **2** $\dfrac{1}{3}$ **3** $\dfrac{1}{4}$ **4** $\dfrac{1}{6}$ **5** $\dfrac{1}{8}$

6 Manolita wants to share this sandwich with 7 friends.
Draw lines to show where the sandwich should be cut
to make 8 pieces of equal length.

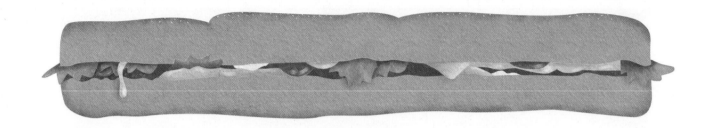

7 Decorate $\frac{1}{2}$ of this strip with dots.

8 Decorate $\frac{1}{4}$ of this strip with stripes.

9 Decorate $\frac{1}{8}$ of this strip with zigzags.

LESSON 7.5 **Fractions of a Set**

Key Ideas

$\frac{1}{4}$ of this group of apples is red. $\frac{1}{4}$ of this group of apples is red.

The fraction $\frac{1}{4}$ tells how many fourths are red,
not how many apples are red.

What fraction of each set is colored?

1 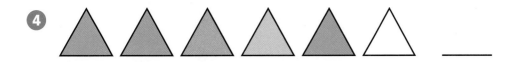 _____

2 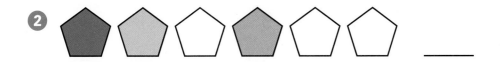 _____

3  _____

4 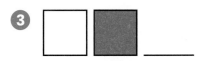 _____

ⓔ Textbook This lesson is available in the *eTextbook.*

5 If 4 tickets are $\frac{1}{2}$ of a set, how many are in a full set? _____

6 If 9 books are $\frac{3}{4}$ of a set, how many are in the whole set? _____

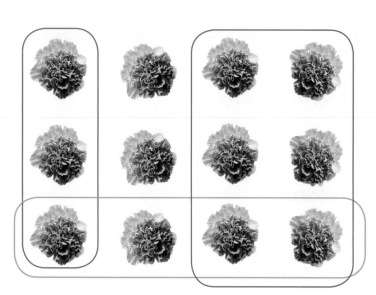

7 What fraction of the set is in the red ring? _____

8 What fraction of the set is in the green ring? _____

9 What fraction of the set is in the blue ring? _____

LESSON 7.6 Fifths

Key Ideas

These pictures show fifths.

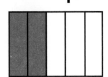 

$\frac{2}{5}$ is red. $\frac{3}{5}$ is green.

Match the fractions that are equal. Then write the fraction for each pair. The first one is done for you.

1

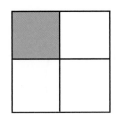   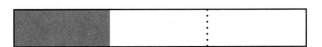

____

2

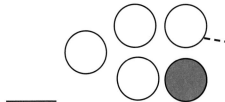

____

3

____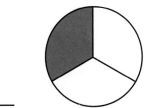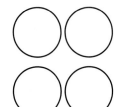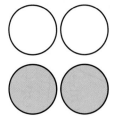

e **Textbook** This lesson is available in the *eTextbook*.

Game

Fraction Game

Game Use these circles to play the **Fraction Game.**

$\frac{1}{1}$

$\frac{1}{2}$ $\frac{1}{2}$ — $\frac{1}{2}$ $\frac{1}{2}$ — $\frac{1}{2}$ $\frac{1}{2}$ — $\frac{1}{2}$ $\frac{1}{2}$

$\frac{1}{3}$ $\frac{1}{3}$ $\frac{1}{3}$ — $\frac{1}{3}$ $\frac{1}{3}$ $\frac{1}{3}$ — $\frac{1}{3}$ $\frac{1}{3}$ $\frac{1}{3}$ — $\frac{1}{3}$ $\frac{1}{3}$ $\frac{1}{3}$

$\frac{1}{4}$ $\frac{1}{4}$ $\frac{1}{4}$ $\frac{1}{4}$ — $\frac{1}{4}$ $\frac{1}{4}$ $\frac{1}{4}$ $\frac{1}{4}$ — $\frac{1}{4}$ $\frac{1}{4}$ $\frac{1}{4}$ $\frac{1}{4}$ — $\frac{1}{4}$ $\frac{1}{4}$ $\frac{1}{4}$ $\frac{1}{4}$

$\frac{1}{5}$ $\frac{1}{5}$ $\frac{1}{5}$ $\frac{1}{5}$ $\frac{1}{5}$ — $\frac{1}{5}$ $\frac{1}{5}$ $\frac{1}{5}$ $\frac{1}{5}$ $\frac{1}{5}$ — $\frac{1}{5}$ $\frac{1}{5}$ $\frac{1}{5}$ $\frac{1}{5}$ $\frac{1}{5}$ — $\frac{1}{5}$ $\frac{1}{5}$ $\frac{1}{5}$ $\frac{1}{5}$ $\frac{1}{5}$

ⓔ Games This game is available as an *eGame.*

Name _____ Date _____

Listen to the problem.

Tim drew a picture and made a physical model to create his design.

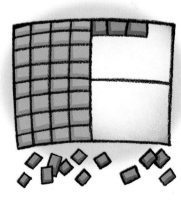

Mila is also making a physical model to create her design.

Show how you solved the problem.

Cumulative Review

Name _____ Date _____

Near Doubles Lesson 2.5

Figure out how much money you
need to buy the listed item or items.

❶ painting book and paints _____

❷ artist palette and painting book _____

❸ colored pencils and sketch book _____

Subtracting Tens Lesson 6.2

Solve this problem.

❹ Juan has $30. The video game costs $60.
How much more money does he need to buy
the video game? He needs _____ dollars.

Chain Calculations Lesson 3.7

Leanna went to a science museum.
She was really excited about the
following items in the gift shop.

❺ How much for
2 molecule models
and 1 heart model? _____

❻ How much would it
cost to buy 3 shells? _____

Molecule model	$2 each
Heart model	$9 each
Shells	$6 each

❼ How much for 1 shell and 1 heart model? _____

❽ How much money would it
cost to buy 3 molecule models? _____

Cumulative Review

Halves and Fourths Lesson 7.1

Answer the questions.

9 The far left edges of the wings on each plane are lined up. Are the halfway marks lined up? _____

10 **Extended Response** Why or why not?

_____

_____

Sixths and Eighths Lesson 7.4

Solve.

11 The 6 gorillas at the zoo are getting pieces of sugar cane plant. Draw lines to show where the keeper should cut the cane to make 6 equal parts.

More Adding Two-Digit Numbers Lesson 5.5

Add.

12
$$\begin{array}{r} 18 \\ + 12 \\ \hline \end{array}$$

13
$$\begin{array}{r} 19 \\ + 21 \\ \hline \end{array}$$

14
$$\begin{array}{r} 21 \\ + 19 \\ \hline \end{array}$$

15
$$\begin{array}{r} 16 \\ + 24 \\ \hline \end{array}$$

16
$$\begin{array}{r} 37 \\ + 33 \\ \hline \end{array}$$

17
$$\begin{array}{r} 30 \\ + 20 \\ \hline \end{array}$$

LESSON 7.7 Equivalent Fractions

Key Ideas

A whole figure can be divided into fractions in many ways.

All the fraction strips below are the same length but are divided into different parts.

Write a fraction on each part of the fraction strips.

1

2

3

4 Shade $\frac{2}{4}$.

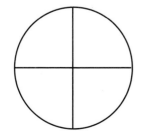

5 Shade $\frac{1}{2}$.

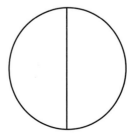

eTextbook This lesson is available in the *eTextbook.*

Answer using pictures and words.

6 Which is bigger: $\frac{1}{2}$ of a pie or $\frac{5}{8}$ of a pie? _____

7 Which is farther: $\frac{1}{3}$ of a mile or $\frac{2}{6}$ of a mile? _____

8 Juanita and Jorge each had six blobs of paint. Draw them.

9 Juanita used $\frac{4}{6}$ of hers. Ring how many she used.

10 Jorge used $\frac{2}{3}$ of his. Ring how many he used.

11 Who used more blobs of paint? _____

LESSON 7.8

Fractions of Numbers

Key Ideas

You can find fractions of numbers just as you can find fractions of objects or sets.

There are 10 craft sticks in a bundle. There are 6 bundles, so there are 60 craft sticks.

$\frac{2}{6}$ of the bundles are red.

20 sticks are red.

$\frac{2}{6}$ of 60 is 20.

$\frac{4}{6}$ of the bundles are blue.

40 sticks are blue.

$\frac{4}{6}$ of 60 is 40.

Find the answers. You may use craft sticks or other objects to help.

1. $\frac{1}{4}$ of 60 = _____

2. $\frac{2}{4}$ of 60 = _____

3. $\frac{3}{4}$ of 60 = _____

4. $\frac{4}{4}$ of 60 = _____

5. $\frac{1}{2}$ of 60 = _____

6. $\frac{2}{2}$ of 60 = _____

7. $\frac{1}{2}$ of 100 = _____

8. $\frac{2}{2}$ of 100 = _____

ⓔ Textbook This lesson is available in the *eTextbook*.

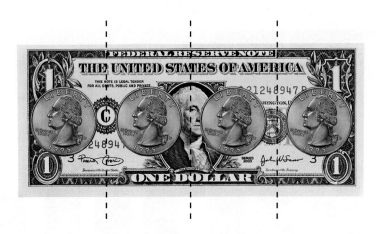

Remember, I whole dollar is worth 100 cents.

$\frac{1}{4}$ of a dollar is worth 25 cents.

$\frac{2}{4}$ of a dollar is worth 50 cents.

9 $\frac{1}{2}$ of 100 = _____

10 $\frac{1}{4}$ of 100 = _____

11 $\frac{2}{4}$ of 100 = _____

12 $\frac{3}{4}$ of 100 = _____

13 $\frac{4}{4}$ of 100 = _____

14 $\frac{1}{2}$ of 50 = _____

LESSON 7.9 **Telling Time—Hour and Half Hour**

Key Ideas

7:00

7:30 or half past 7

What time is it? Write your answers.

1  _____

2  _____

3  _____

4  _____

5  _____

6  _____

7  _____

8  _____

eTextbook This lesson is available in the *eTextbook*.

Show the time.

9 4:00

12 six o'clock

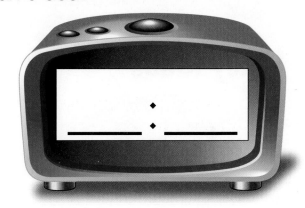

10 ten thirty

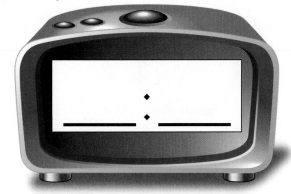

13 7:30

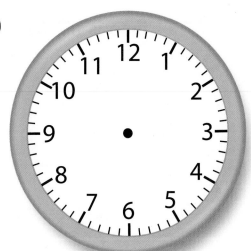

11 1:30

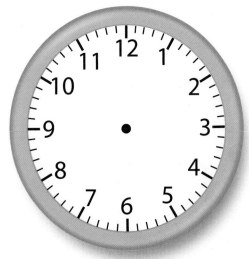

14 3:30

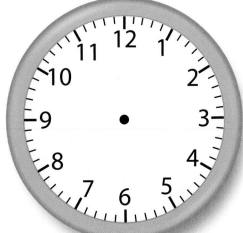

Name _____ Date _____

LESSON
7.10

Telling Time—Quarter Hour

Key Ideas

There are different ways to say and write
the same time.

7 o'clock

7:00

quarter past 7

7:15

half past 7

7:30

quarter to 8

7:45

What time is it?

1  _____ o'clock

_____ : _____

2  half past _____

_____ : _____

3  quarter to _____

_____ : _____

4  half past _____

_____ : _____

What time is it?

5

quarter after _____

_____ : _____

8

quarter to _____

_____ : _____

6

_____ o'clock

_____ : _____

9

half past _____

_____ : _____

7

_____ o'clock

_____ : _____

10

quarter after _____

_____ : _____

LESSON 7.11 **Fractions Greater than One**

Key Ideas

Fractions represent parts of a whole.
If you put together enough fractions,
you can have more than 1 whole.

The Museum Café sells pie by the slice.
The afternoon tour group bought 9 slices.

When they put them together, they can see
they bought more than 1 whole pie. They bought
$1\frac{1}{2}$ pies. What fraction of the pie is one slice?

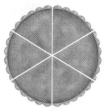

Look at the fraction. If it is more than 1 whole,
rewrite it in a different form. The first one is
done for you.

1   $\frac{6}{4}$ (More) or Less $1\frac{2}{4}$ or $1\frac{1}{2}$ _____

2   $\frac{3}{2}$ More or Less   _____

3   $\frac{5}{8}$ More or Less   _____

4   $\frac{15}{12}$ More or Less   _____

Juice boxes come in packages of 8.

5 What fraction of the whole
package is 1 juice box? _____

6 Juice boxes come in packages of 8.
Write the fraction for the juice boxes below.

_____

7 How many whole packages of juice
boxes are there in Problem 6? _____

8 How many juice boxes are left over? _____

Draw circles to represent these fractions.

9 $\frac{3}{2}$

10 $\frac{6}{4}$

Name _____ Date _____

Listen to the problem.

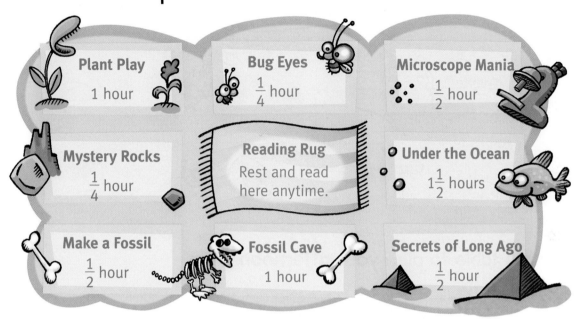

Plant Play 1 hour

Bug Eyes $\frac{1}{4}$ hour

Microscope Mania $\frac{1}{2}$ hour

Mystery Rocks $\frac{1}{4}$ hour

Reading Rug Rest and read here anytime.

Under the Ocean $1\frac{1}{2}$ hours

Make a Fossil $\frac{1}{2}$ hour

Fossil Cave 1 hour

Secrets of Long Ago $\frac{1}{2}$ hour

Use the table to plan your morning at the museum.

Name of Activity	Starting Time	How Long	Ending Time
	9:00		

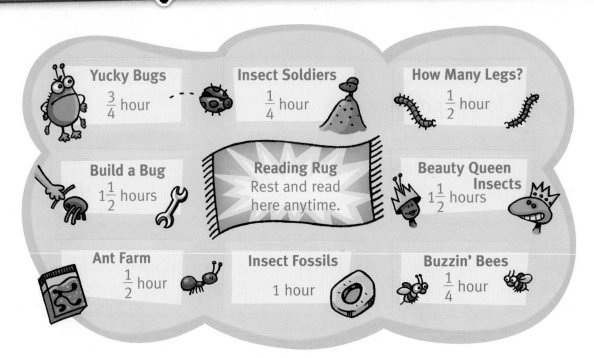

Yucky Bugs
$\frac{3}{4}$ hour

Insect Soldiers
$\frac{1}{4}$ hour

How Many Legs?
$\frac{1}{2}$ hour

Build a Bug
$1\frac{1}{2}$ hours

Reading Rug
Rest and read here anytime.

Beauty Queen Insects
$1\frac{1}{2}$ hours

Ant Farm
$\frac{1}{2}$ hour

Insect Fossils
1 hour

Buzzin' Bees
$\frac{1}{4}$ hour

Use the table to plan your afternoon from 12:30 to 2:45.

Name of Activity	Starting Time	How Long	Ending Time
	12:30		

Cumulative Review

Name _____ Date _____

Tally Marks Grade 1 Lesson 2.4

Write how many tally marks.

❶ Sierra made a survey of the 70 kids at summer camp.
She asked each camper which activity he or she enjoyed.
Write the count of each set of tally marks.

Activity	How Many	Number
Exploring wetlands	卌 卌 卌 I	
Hiking	卌 III	
Nature stories	卌 卌	
Safety skills	卌 卌 I	

Collecting and Recording Data Lesson 4.6

Use the table above to answer these questions.

❷ How many campers enjoyed hiking? _____

❸ How many campers liked safety skills? _____

❹ Which activity was enjoyed by 16 campers? _____

❺ What doesn't the table tell you? _____

Fractions of Numbers Lesson 7.8

Find the answers.

❻ $\frac{2}{4}$ of 100 is _____   ❼ $\frac{1}{2}$ of 50 is _____   ❽ $\frac{3}{4}$ of 100 is _____

Cumulative Review

Fractions of a Set Lesson 7.5

Ring the correct number of sculptures.

9 Ring $\frac{1}{4}$ of the sculptures. How many? _____

10 Ring $\frac{1}{2}$ of the sculptures. How many? _____

11 Ring $\frac{1}{3}$ of the sculptures. How many? _____

- -

Telling Time–Quarter Hour Lesson 7.10

Write the time.

12 quarter to _____

_____ : _____

13 half past _____

_____ : _____

14 quarter after _____

_____ : _____

15 _____ o'clock

_____ : _____

Real Math • Chapter 7

Name _____ Date _____

In this chapter you learned about fractions. You learned to identify fractions of shapes, sets of objects, and numbers. You also learned about using fractions to tell time on a clock.

· ·

Color the correct fraction of each figure.

1 $\frac{1}{2}$ ◯

2 $\frac{1}{3}$ ▢

3 $\frac{1}{4}$ ▢

4 $\frac{2}{4}$ ▢

Follow the directions.

5 This line represents $\frac{1}{2}$ of a road. Draw the rest of the road.

▬▬▬▬▬▬▬

Follow the directions.

6 These crayons are $\frac{1}{2}$ of a set. Draw the rest of the set.

7 Ring $\frac{1}{3}$ of the starfish.

How many? _____

Ring the letter of the correct answer.

8 What fraction of 60 is 15?

   a. $\frac{1}{2}$ **b.** $\frac{1}{3}$ **c.** $\frac{1}{4}$

9 Which coin is worth $\frac{1}{4}$ of a dollar?

   a.   **b.**   **c.**

10 What is another way to say and write 7:15?

   a. quarter past 7

   b. half past 7

   c. quarter to 8

Name _____ Date _____

Lesson 7.5 **Solve.**

1 If 6 people are $\frac{3}{4}$ of a tour group, how many are in the whole group? _____

Lessons 7.1, 7.3 **Read** the story. Draw your answers on the picture.

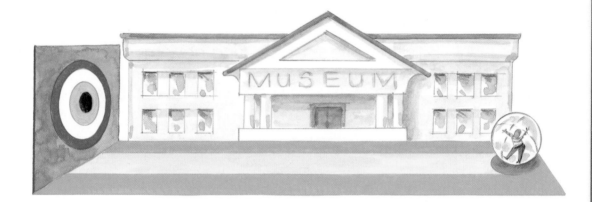

2 Raul and Jonah were at the science museum. They each tried walking inside a ball. Raul made it one-fourth of the way along the path. Draw an *R* where he stopped. Jonah made it three-fourths of the way. Draw a *J* where he stopped.

Lessons 7.4, 7.6 **Write** the fraction that represents the shaded part.

3 _____ **4** _____

Lessons 7.7–7.8 **Find** the answers.

The three bears—Growly, Brutus, and Bambam—each had six apples.

5 Growly ate $\frac{4}{6}$ of his. Ring how many he ate.

6 Brutus ate $\frac{1}{3}$ of his. Ring how many he ate.

7 Bambam ate $\frac{1}{2}$ of his. Ring how many he ate.

8 $\frac{1}{2}$ of 60 = _____ $\frac{2}{4}$ of 60 = _____

Lessons 7.9–7.10 **What** time is it? Write or draw your answers.

9

Four o'clock

10

_____

Name _____ Date _____

Write the fraction that is shaded.

1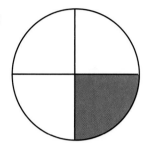

3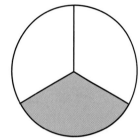

2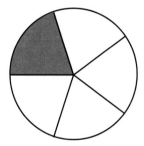

4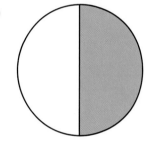

Copy each figure. Color each figure to match the fraction.

5 $\frac{1}{6}$

7 $\frac{1}{4}$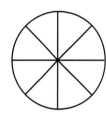

6 $\frac{5}{8}$

8 $\frac{3}{5}$

What fraction is shown by the shaded part?

9

a. $\frac{3}{4}$　　b. $\frac{1}{3}$　　c. $\frac{3}{3}$　　d. $\frac{4}{4}$

11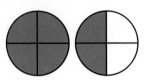

a. $\frac{1}{2}$　　b. $\frac{2}{4}$　　c. $\frac{3}{4}$　　d. $\frac{6}{4}$

10

a. $\frac{1}{2}$　　b. $\frac{1}{3}$　　c. $\frac{1}{6}$　　d. $\frac{3}{6}$

12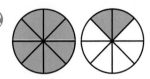

a. $\frac{1}{4}$　　b. $\frac{3}{8}$　　c. $\frac{10}{8}$　　d. $\frac{6}{4}$

What time is it?

13

a. quarter after 4

b. quarter to 5

c. half past 4

d. half past 5

14

a. quarter after 2

b. half past 2

c. quarter to 2

d. quarter to 3

Name _____ Date _____

What time is it?

15
a. 11:00
b. 12:00
c. 11:30
d. 12:30

16
a. 9:15
b. 10:15
c. 9:30
d. 9:45

Count by tens. Add or subtract.

17 90 − 20 =
a. 80   b. 70
c. 60   d. 50

18 50 − 40 =
a. 90   b. 30
c. 10   d. 0

19 20 + 40 =
a. 20   b. 30
c. 50   d. 60

Solve.

20 Morris spent $48 for a jacket and $25 for a pair of jeans. How much did Morris spend in all?

a. $83   b. $73   c. $63   d. $23

21 Deshawn sold 39 tickets for the school carnival. Jarrell sold 57 tickets. How many more tickets did Jarrell sell than Deshawn?

a. 96   b. 22   c. 18   d. 12

22 Penny rode her bicycle 25 kilometers on Friday and 17 kilometers on Saturday. How many kilometers did she ride?

a. 42   b. 32   c. 12   d. 8

Practice Test

Draw pictures to show the fractions.

23. Draw a rectangle to show eighths. Write the fraction for each part. Then color $\frac{1}{4}$ of the rectangle.

24. Draw a rectangle and color $\frac{1}{2}$. Can you use this rectangle to show thirds? Explain.

____

____

25. Draw fraction circles to show $\frac{2}{6}$, $\frac{1}{3}$, and $\frac{2}{4}$. Then draw a ring around the fraction that shows the most. Draw an X on the two fractions that are equal.

Thinking Story

Half a Job

Draw a job that you could do in two days if you did half the job each day.

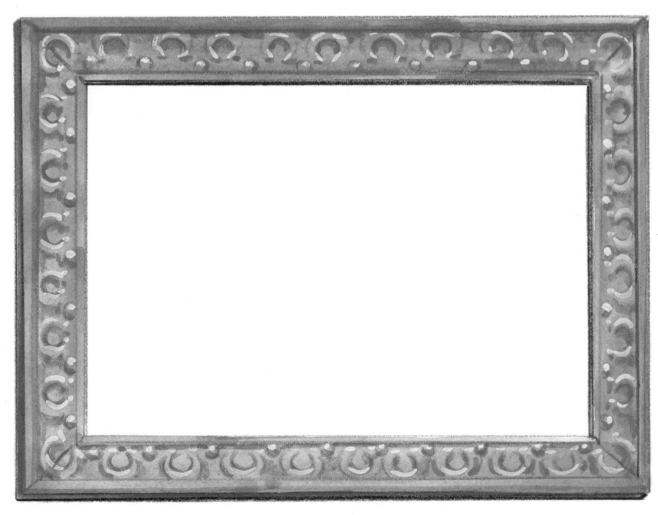

① **Color** half the big square with blue.

② **Color** half of what is left with red.

③ **Color** half of what is left with green.

Real Math • Chapter 7

In This Chapter You Will Learn

- about plane figures and space figures.
- how to identify obtuse, acute, and right angles.
- about congruence and symmetry.

Name _____ Date _____

Trace and cut out the track. Use it as a model to create many tracks with different characteristics.

LESSON 8.1 **Tessellations**

Key Ideas

These are examples of plane figures.

 hexagon   triangle   square   rhombus   trapezoid

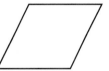

Tessellations are patterns of plane figures that do not have any spaces or gaps between them.

Ring the shapes that make up each pattern.

1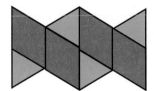

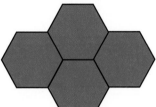

2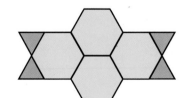

3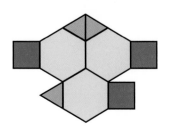

④ **Use** these shapes to fill the hexagons. Can you fill them six different ways?

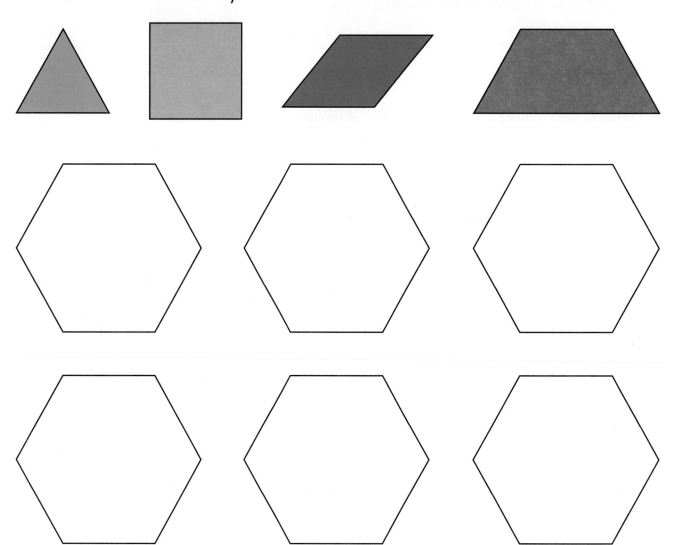

⑤ Which shape could not be used for making one of the hexagons? _____

LESSON 8.2 **Parts of Plane Figures**

Key Ideas

The smaller shapes have area that is a fraction of the area of the larger shapes.

Draw a ring around the correct fraction.
Use pattern blocks to help.

1
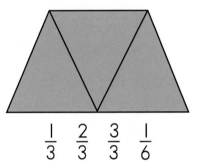

$\frac{1}{3}$ $\frac{2}{3}$ $\frac{3}{3}$ $\frac{1}{6}$

4
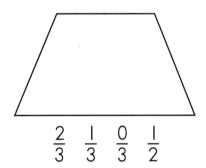

$\frac{2}{3}$ $\frac{1}{3}$ $\frac{0}{3}$ $\frac{1}{2}$

2
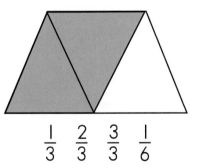

$\frac{1}{3}$ $\frac{2}{3}$ $\frac{3}{3}$ $\frac{1}{6}$

5
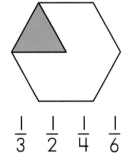

$\frac{1}{3}$ $\frac{1}{2}$ $\frac{1}{4}$ $\frac{1}{6}$

3
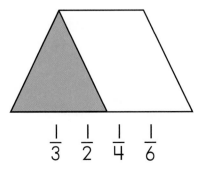

$\frac{1}{3}$ $\frac{1}{2}$ $\frac{1}{4}$ $\frac{1}{6}$

6
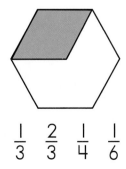

$\frac{1}{3}$ $\frac{2}{3}$ $\frac{1}{4}$ $\frac{1}{6}$

Textbook This lesson is available in the *eTextbook*.

Draw a ring around the correct fraction.
Use pattern blocks to help.

7

$\dfrac{2}{6}$ $\dfrac{2}{3}$ $\dfrac{3}{3}$ $\dfrac{1}{6}$

11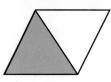

$\dfrac{1}{3}$ $\dfrac{1}{2}$ $\dfrac{1}{4}$ $\dfrac{1}{6}$

8

$\dfrac{1}{3}$ $\dfrac{2}{3}$ $\dfrac{3}{3}$ $\dfrac{4}{6}$

12

$\dfrac{1}{3}$ $\dfrac{3}{6}$ $\dfrac{3}{3}$ $\dfrac{1}{6}$

9

$\dfrac{2}{6}$ $\dfrac{2}{3}$ $\dfrac{3}{3}$ $\dfrac{1}{6}$

13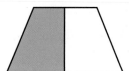

$\dfrac{1}{3}$ $\dfrac{1}{2}$ $\dfrac{1}{4}$ $\dfrac{2}{3}$

10

$\dfrac{1}{2}$ $\dfrac{1}{3}$ $\dfrac{1}{4}$ $\dfrac{1}{5}$

14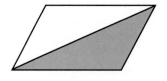

$\dfrac{1}{3}$ $\dfrac{1}{2}$ $\dfrac{1}{4}$ $\dfrac{1}{6}$

LESSON 8.3 Combining Plane Figures

Key Ideas

Larger shapes are often made of smaller shapes.

Use pattern blocks to fill these shapes.
Use the blocks shown.

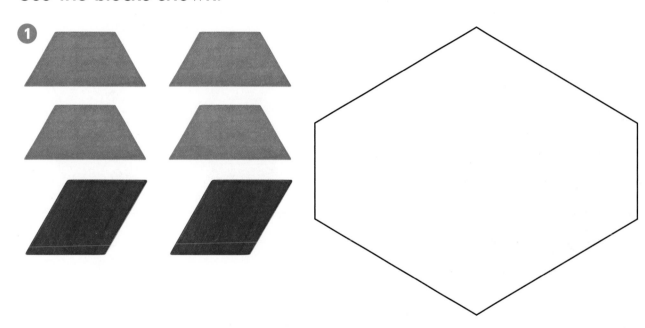

1

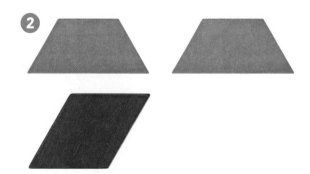

2

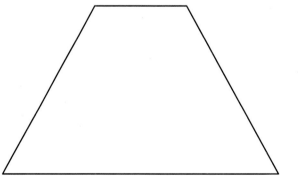

Use pattern blocks to fill these shapes.
Use the blocks shown.

3

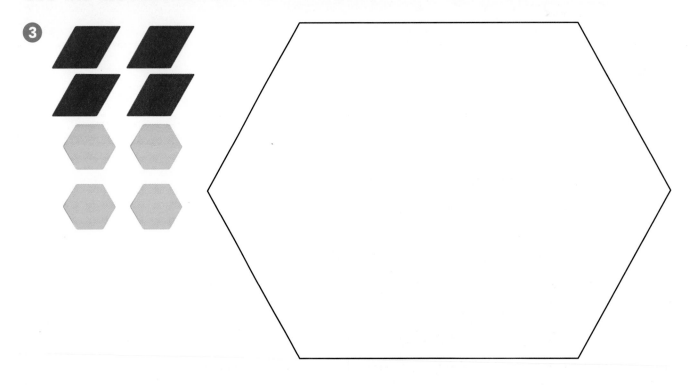

4

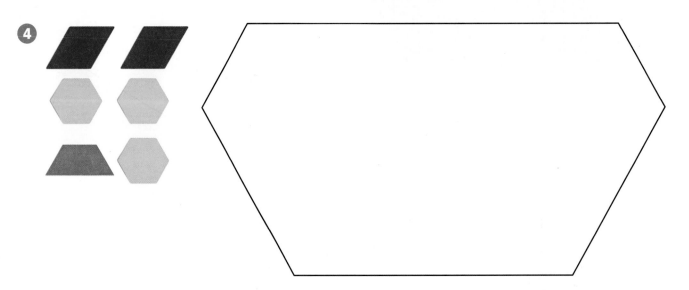

LESSON 8.4 Defining Plane Figures—Quadrilaterals

Key Ideas

Figures with four sides are called quadrilaterals.

This is a quadrilateral.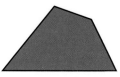

These are special types of quadrilaterals.

Trapezoid   Parallelogram   Rhombus

Rectangle   Square

1 Draw a line to make two rectangles.

2 Draw a line to make two parallelograms.

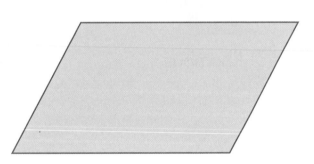

3 Draw a line differently to make two parallelograms.

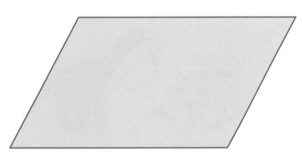

4 Draw a line to make two quadrilaterals.

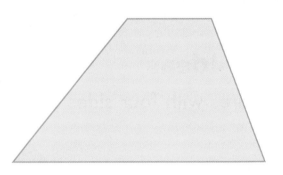

5 Draw a line to make two triangles.

6 Draw lines to make four squares.

Name _____ Date _____

Listen to the problem.

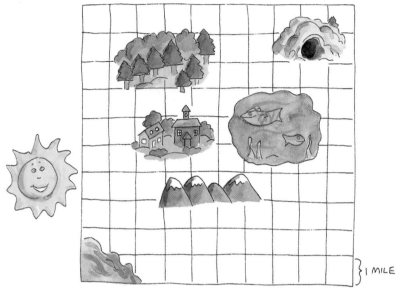

My route back to the village at 3:00:

1. Face the sun, and go parallel to the lake for 2 miles.

2. Make a $\frac{1}{4}$ turn to the right.

3. Go 2 miles.

4. Make a $\frac{1}{4}$ turn to the left.

5. Go 3 miles.

6. Make a $\frac{1}{4}$ turn to the left so that you are facing the mountains.

7. Walk 3 miles.

8. Arrive at the village.

Mitch is using guess, check, and revise to solve the problem.

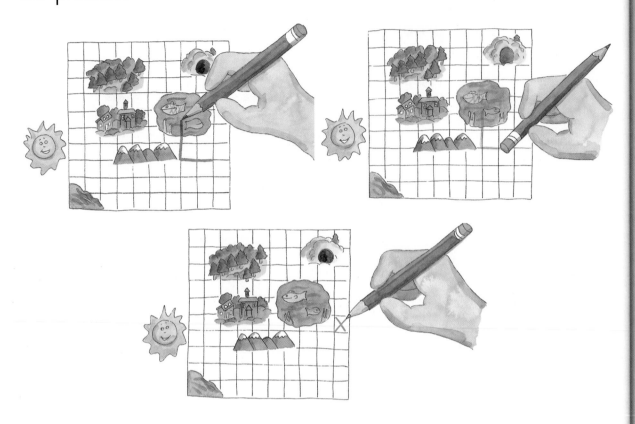

Gwen is working backward to solve the problem.

Cumulative Review

Name _____ Date _____

Pictographs Lesson 4.9

Use the pictograph to answer the questions. Each picture of a balloon stands for 10 balloons.

Mrs. Paul	🎈	🎈	🎈	
Mia	🎈	🎈	🎈	🎈
Alex	🎈			

❶ Who blew up the fewest balloons? _____

How many? _____

❷ Who blew up the most balloons? _____

How many? _____

❸ How many did Mrs. Paul blow up? _____

Adding Two-Digit Numbers Lesson 5.4

Solve.

❹ 73 + 18 = _____   ❼ 31 + 29 = _____

❺ 41 + 28 = _____   ❽ 13 + 68 = _____

❻ 86 + 69 = _____   ❾ 19 + 11 = _____

Cumulative Review

Combining Plane Figures Lesson 8.3

Use pattern blocks to fill in the outline.

⑩ Use 4 red trapezoids and 2 blue rhombuses to make a hexagon.

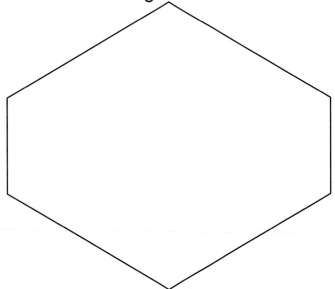

· ·

Subtraction Facts Lesson 3.5

Subtract.

⑪ 17 − 8 = _____   ⑮ 9 − 5 = _____

⑫ 10 − 3 = _____   ⑯ 15 − 10 = _____

⑬ 10 − 8 = _____   ⑰ 16 − 8 = _____

⑭ 10 − 6 = _____   ⑱ 13 − 9 = _____

LESSON 8.5 Obtuse, Acute, and Right Angles

Key Ideas

Three types of angles are obtuse, acute, and right.

Acute angle

Right angle

Obtuse angle

Write the correct time. Then write if the measure of the angle is right, acute, or obtuse.

1  _____
_____

4  _____
_____

2  _____
_____

5  _____
_____

3  _____
_____

6  _____
_____

All of these are right angles.

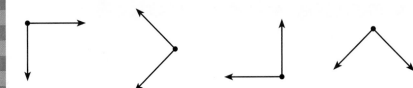

None of these is a right angle.

7 **Ring** the right angles.

8 **Identify** the angles.

a.  _____

b. _____

c. _____

d. _____

Name _____   Date _____

LESSON 8.6 **Congruence**

Key Ideas

Two figures that are the same shape and the same size are called congruent.

Here are some ways to describe the movement of a figure.

Slide 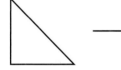 Flip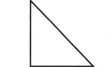

Turn

Trace the first figure onto paper, and then slide, turn, or flip the paper to find the congruent figure in each row. Ring the congruent figure.

1

2

Copyright © SRA/McGraw-Hill.

ⓔ Textbook This lesson is available in the *eTextbook*.   301

Ring the figure that shows what the first figure
will look like after you slide, turn, or flip it.

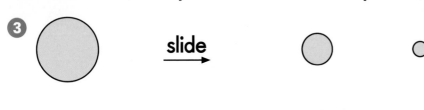

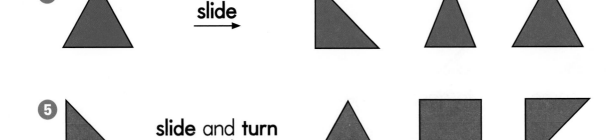

Writing + Math **Journal**

Can you think of a time when it would be important to
have objects congruent to one another? Why is it important
for these objects to be congruent?

LESSON 8.7 Symmetry

Key Ideas

If you can hold a mirror along a line through a polygon and have the figure look the same with and without the mirror, that line is called a line of symmetry.

Hold a mirror to each figure, and look for the lines of symmetry. Then draw one of the lines you find.

1

4
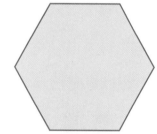

2
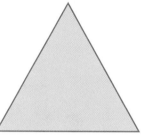

5

3

6

Draw lines of symmetry. If a letter has no lines of symmetry, draw a ring around it.

7    8   9    10

11   12    13    14

15   16   17   18

19    20   21   22

23    24    25    26

27    28    29

30   31   32

Real Math • Chapter 8 • Lesson 7

LESSON 8.8

Defining Other Plane Figures

Key Ideas

Polygons are closed figures with at least three straight sides. Polygons are named by counting the number of sides. Here are some examples.

Right triangle
(3 sides and a right angle)

Triangle
(3 sides)

Quadrilateral
(4 sides)

Pentagon
(5 sides)

Hexagon
(6 sides)

Octagon
(8 sides)

A **regular polygon** has sides that are all the same length and angles that are all equal.

This is a regular pentagon.

These are irregular pentagons.

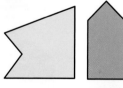

This is a circle. A circle is not a polygon.

Follow the directions.

① **Draw** a circle.

② **Draw** a right triangle.

③ **Draw** a regular triangle.

④ **Draw** a triangle with one obtuse angle.

⑤ **Draw** an irregular hexagon.

⑥ **Draw** a different irregular hexagon.

LESSON 8.9 Space Figures

Key Ideas

Space figures cannot fit in a plane. Here are some examples of space figures.

Cone

Pyramid

Sphere

Cylinder

Cube

Rectangular prism

e Textbook This lesson is available in the *eTextbook*.

Match each figure to the correct name.

Pyramid

Sphere

Rectangular prism

Cube

Cylinder

Find the space figures in this picture. Ring the pictures when you find them. Write the names of the figures you find.

_____

_____

_____

_____

_____

Cut out each figure. Fold and tape to make two different pyramids.

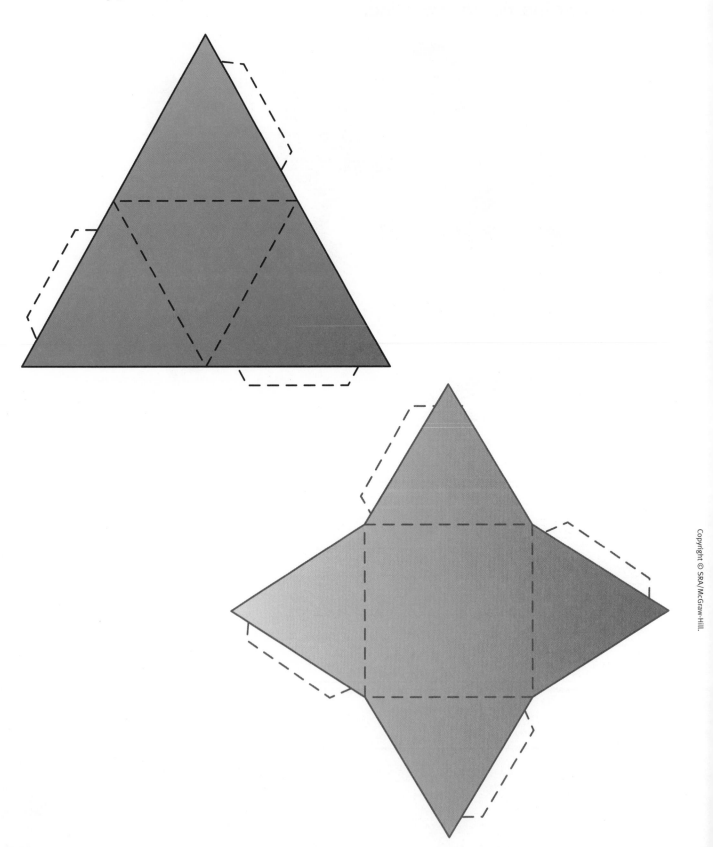

LESSON 8.10 Space Figures and Plane Figures

Key Ideas

You can compare space figures and plane figures.

1 Study the plane figures. Complete the table.

Plane Figure	Number of Sides	Number of Corners
Triangle		
Square		
Rectangle		
Parallelogram		
Pentagon		

 **Journal**

Is it possible to draw a polygon with a number of corners that is not equal to the number of sides? Explain.

Copyright © SRA/McGraw-Hill.

2 Study the space figures. Complete the table.

Space Figure	Number of Flat Faces	Number of Edges	Number of Vertices
Cube			
Square pyramid			
Rectangular prism			
Cylinder			

3 Draw a net for a rectangular prism.

4 What space figure do you think can be made from this net?

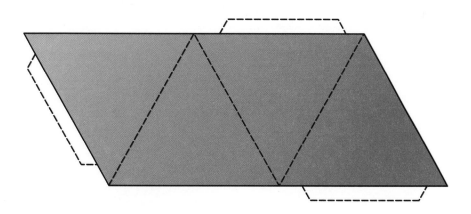

Different nets can make a cube.

6 Draw a net for a cube.

7 Draw a different net for a cube.

Name _____ Date _____

Listen to the problem.

Draw the fossil you made.

Draw a plan for a box for your fossil.

Real Math • Chapter 8 • Exploring Problem Solving

Cumulative Review

Name _____ Date _____

Telling Time—Quarter Hour Lesson 7.10

Write each time two different ways.

1    **2**    **3**

half past ____   ____ o'clock   quarter past ____

____ : _____   ____ : _____   ____ : _____

Obtuse, Acute, and Right Angles Lesson 8.5

Ring the type of angle shown for each of the clocks in Problems 1–3.

4 Orange clock: acute   obtuse   right

5 Green clock: acute   obtuse   right

6 Blue clock: acute   obtuse   right

Remaining Facts and Function Machines Lesson 2.6

Add.

7 9 + 9 = _____   **8** 3 + 7 = _____   **9** 10 + 10 = _____

10 6 + 8 = _____   **11** 1 + 5 = _____   **12** 7 + 2 = _____

Cumulative Review

Addition and Subtraction Functions Lesson 3.3

Fill in the missing numbers.

⑬

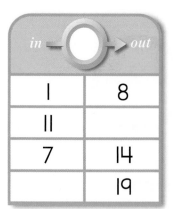

in	out
1	8
11	
7	14
	19

The rule is _____.

⑭

in	out
65	52
43	30
20	7
16	3

The rule is _____.

Place Value Lesson 5.1

How many sticks? Write your answers.

⑮  _____

⑯ _____

Subtracting Tens Lesson 6.2

Subtract.

⑰ 40 − 10 = _____

⑱ 70 − 20 = _____

Name _____ Date _____

In this chapter you learned about geometry.
You learned about plane figures and space
figures. You learned to identify angles.

....

Ring the letter of the correct answer.

1 Which of these figures is not a quadrilateral?

a.   b. 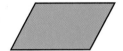   c.   d.

2 Which of these is a right angle?

a.   b.    c.   d.

3 Which triangle below is congruent to this triangle?

a.   b.    c.   d.

Follow the directions.

4 Draw a line to show how this square
can be divided into two triangles.

5 Draw a line of symmetry
through each of these letters. A E I O U

Follow the directions.

6 Draw a 5-sided polygon.

7 Ring the space figure.

Write your answers.

This is a rectangular prism.

8 How many faces? _____

9 How many edges? _____

10 How many vertices? _____

Name _____ Date _____

Lesson 8.5 **Look** at the opening of the cracked dinosaur egg. Ring the correct angle of the opening.

Obtuse Acute Right   Obtuse Acute Right   Obtuse Acute Right

Lesson 8.1 **Ring** the shapes that make up each pattern.

4

5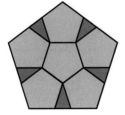

Lesson 8.4 **Draw** a line on each figure to make the new figures.

6 Make two parallelograms.

8 Make two right triangles.

7 Make two quadrilaterals.

9 Make two pentagons.

Lesson 8.2 **What** fraction of the area of each figure is filled? Draw a ring around the letter of the correct fraction.

10

a. $\dfrac{1}{3}$

b. $\dfrac{1}{2}$

c. $\dfrac{1}{4}$

d. $\dfrac{1}{6}$

11

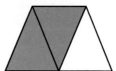

a. $\dfrac{1}{3}$

b. $\dfrac{2}{3}$

c. $\dfrac{3}{3}$

d. $\dfrac{1}{6}$

Practice Test

Name _____ Date _____

Write the fraction that names the shaded part.

1  _____

3  _____

2 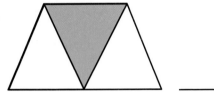 _____

4 _____

Ring the pairs of shapes that are congruent.

5

8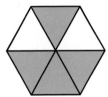

6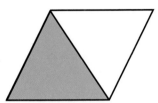

7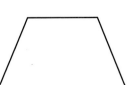

e Textbook This lesson is available in the **eTextbook**.

Practice Test

Identify each figure.

9

a. hexagon

b. trapezoid

c. rhombus

d. square

10

a. cube

b. cylinder

c. rectangular prism

d. pyramid

11

a. sphere

b. pyramid

c. cube

d. cylinder

Identify each angle.

12

a   acute

b. obtuse

c. right

d. straight

13

a. acute

b. obtuse

c. right

d. straight

14

a   acute

b. obtuse

c. right

d. straight

Ring the letter of the correct answer.

15 How many edges are on a cube?

a. 0   b. 6

c. 8   d. 12

16 How many vertices are on a cylinder?

a. 0   b. 6

c. 8   d. 12

Practice Test

Name _____ Date _____

What time is shown?

17

a. quarter to 5

b. quarter to 6

c. quarter after 3

d. quarter after 5

18

a. 4 o'clock

b. 8 o'clock

c. half past 4

d. half past 8

19

a. 8:00

b. 9:00

c. 8:30

d. 9:30

Write the fraction that matches the shaded part.

20

a. $\frac{1}{2}$ b. $\frac{1}{3}$

c. $\frac{1}{4}$ d. $\frac{4}{4}$

22

a. $\frac{1}{2}$ b. $\frac{2}{2}$

c. $\frac{3}{2}$ d. $\frac{3}{3}$

21

a. $\frac{1}{6}$ b. $\frac{2}{6}$

c. $\frac{2}{2}$ d. $\frac{1}{2}$

23

a. $\frac{8}{6}$ b. $\frac{6}{6}$

c. $\frac{6}{8}$ d. $\frac{2}{6}$

Use pattern blocks.

24 Use pattern blocks to make a large triangle. Draw blocks to fill the figure.

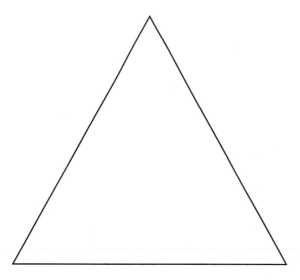

Make a drawing.

25 Draw a square. Then draw as many lines of symmetry as possible in the square.

Name _____   Date _____

Manolita Changes Things

Manolita has 20 pictures and 5 sheets of construction paper. Show how she would have to cut the paper to make a book with 1 picture on each piece.

Ring five things that are changed in the
second picture.

Real Math • Chapter 8

CHAPTER 9
Three-Digit Addition and Subtraction

In This Chapter You Will Learn
- how to add and subtract three-digit numbers.
- place value for five-digit numbers.

Name _____ Date _____

Listen to the problem.

We have _____ stamps.

This is how we organized our stamps.

Name _____ Date _____

Key Ideas

In the number 6,254 there are 6 thousands, 2 hundreds, 5 tens, and 4 ones.

thousands	hundreds	tens	ones
6	2	5	4

① In 725

How many hundreds? _____

How many tens? _____

How many ones? _____

② In 384

How many hundreds? _____

How many tens? _____

How many ones? _____

③ In 2,536

How many thousands? _____

How many hundreds? _____

How many tens? _____

How many ones? _____

④ In 4,730

How many thousands? _____

How many hundreds? _____

How many tens? _____

How many ones? _____

⑤ Fill in the place value table with

2 in the tens place,
4 in the ones place, and
7 in the hundreds place.

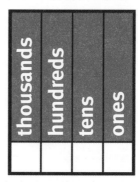

thousands	hundreds	tens	ones

Write the number. _____

Mark a point on the number line where you think each number belongs, and label it.

6 5, 7, 9, 4

0   10

7 17, 99, 50, 80

0   100

8 400; 9,600; 3,000; 7,000

0   10,000

9 4,400; 5,600; 5,000; 100

0   10,000

10 500, 800, 200, 100

0   1,000

11 501, 801, 500, 950, 200

0   1,000

12 400, 960, 300, 700, 100

0   1,000

Writing + Math **Journal**

Use these digits: 1, 2, 3, 4. Write all the possible four-digit numbers. Order them from greatest to least.

LESSON 9.2 Modeling Numbers through 1,000

Key Ideas

10 hundreds equal
1 thousand.

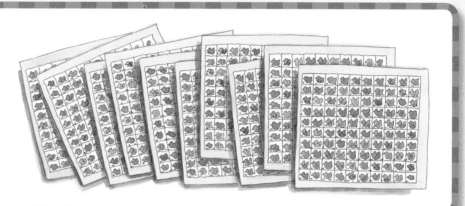

Write the number shown.

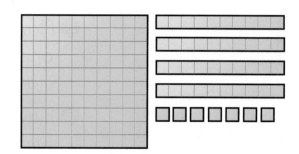

1 _____

2 _____

Draw pictures to show the number.
Instead of drawing 10 or 100 items,
you may show 10 and 100 like this: | 10 | | 100 |

3 253 bells

4 302 leaves

Write the following numbers in words.

5 782 _____

7 94 _____

6 361 _____

8 7,126 _____

Number Sequence and Strategies Practice

Harder Counting and Writing Numbers Game

Players: Two or more

Materials: Paper and pencils

HOW TO PLAY

❶ Player One chooses a starting number and an ending number between 0 and 1,000 (for example, 286 and 325). Make sure the numbers are not more than 100 apart.

❷ Player Two counts on and writes one, two, or three numbers from the starting number.

❸ Player One counts on and writes one, two, or three more numbers.

❹ The players take turns counting on and writing. The player who counts to and writes the ending number wins.

LESSON 9.3 Comparing Numbers through 1,000

Key Ideas

You can use the greater than sign > and less than sign < to show which number is greater. The wider end of the symbol always points to the greater amount. Use an equal sign = if both sides are the same.

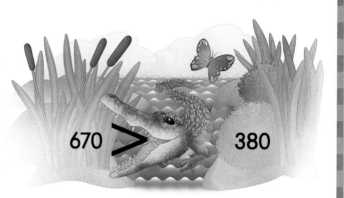

670 > 380

Write how many. Draw the correct sign.

1

_____   _____

2 350 ◯ 370   **3** 450 ◯ 620   **4** 820 ◯ 770

5 560 ◯ 560   **6** 200 ◯ 190   **7** 100 ◯ 1,000

8 400 + 20 ◯ 400 + 10   **10** 750 + 30 ◯ 750 + 40

9 400 − 20 ◯ 400 − 10   **11** 750 − 30 ◯ 750 − 40

Write the words *greater than, less than,*
or *equal to* to make the statements true.

12 Four hundred is _____ three hundred.

13 Two hundred fifty is _____ two hundred sixty.

Ring the letter of the number that matches
the number shown.

14 1,000

    a. one thousand
    b. one hundred
    c. one

15 930

    a. ninety-three
    b. nine hundred
    c. nine hundred thirty

16 Extended Response Is 589 − 15 greater than, less than,
or equal to 588 − 14? How do you know?

_____

_____

Game Play the **Harder Counting and Writing
Numbers Game** in pairs.

 **Writing + Math** **Journal**
Draw a picture to represent a three-digit number such as 457.

LESSON 9.4

Adding Multiples of 10

Key Ideas

You can add three-digit numbers in the same way you add two-digit numbers.

$$\begin{array}{r} 24 \\ + 49 \\ \hline 73 \end{array}$$

and

$$\begin{array}{r} 240 \\ + 490 \\ \hline 730 \end{array}$$

Add.

1 $\begin{array}{r} 29 \\ + 67 \\ \hline \end{array}$

2 $\begin{array}{r} 43 \\ + 18 \\ \hline \end{array}$

3 $\begin{array}{r} 30 \\ + 70 \\ \hline \end{array}$

4 $\begin{array}{r} 32 \\ + 24 \\ \hline \end{array}$

5 $\begin{array}{r} 320 \\ + 240 \\ \hline \end{array}$

6 $\begin{array}{r} 300 \\ + 700 \\ \hline \end{array}$

7 $\begin{array}{r} 430 \\ + 180 \\ \hline \end{array}$

8 $\begin{array}{r} 290 \\ + 670 \\ \hline \end{array}$

Fill in the missing numbers.

9

in → +10 → out	
70	
	110
130	140
160	

10

in → +100 → out	
300	400
	600
750	
900	

Write the number.

11 Four hundred twenty-five _____

12 Nine hundred forty-seven _____

13 Two hundred sixty-three _____

Game Play the **Rummage Sale Game.**

Writing + Math **Journal**

Explain how the addition of two-digit numbers is different from the addition of three-digit numbers.

LESSON 9.5 Three-Digit Addition

Key Ideas

Remember, you can add three-digit numbers in the same way you add two-digit numbers.

$325 + 289 = \underline{\quad ? \quad}$

$$\begin{array}{r} 325 \\ + 289 \\ \hline \end{array}$$

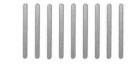

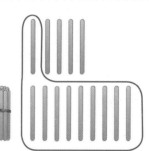

$$\begin{array}{r} 1 \\ 325 \\ + 289 \\ \hline 4 \end{array}$$

$$\begin{array}{r} 11 \\ 325 \\ + 289 \\ \hline 14 \end{array}$$

$$\begin{array}{r} 11 \\ 325 \\ + 289 \\ \hline 614 \end{array}$$

Add.

1   432
    + 158

2   276
    + 395

3   403
    + 608

4   302
    + 708

5   536
    + 492

6   507
    + 394

7   574
    + 687

8   574
    + 688

9   574
    + 689

10 **Extended Response** **Solve.** **Explain** how the two are similar.

    32   320
    + 57   + 570

_____

_____

_____

LESSON 9.6 Subtracting Multiples of 10

Key Ideas

Subtracting three-digit numbers is similar to subtracting two-digit numbers. You may regroup.

$24 - 7 = 17$ is like $240 - 70 = 170$

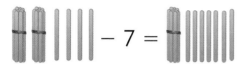 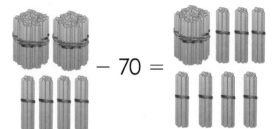

Subtract.

1
```
   15
 − 7
```

2
```
   43
 − 9
```

3
```
   62
 − 5
```

4
```
   14
 − 8
```

5
```
  150
 − 70
```

6
```
  620
 − 50
```

7
```
  430
 − 90
```

8
```
  140
 − 80
```

Subtract.

⑨
$$\begin{array}{r} 14 \\ -\ 6 \\ \hline \end{array}$$

⑩
$$\begin{array}{r} 520 \\ -\ 60 \\ \hline \end{array}$$

⑪
$$\begin{array}{r} 11 \\ -\ 9 \\ \hline \end{array}$$

⑫
$$\begin{array}{r} 140 \\ -\ 60 \\ \hline \end{array}$$

⑬ The regular price for a computer is $943.
How much will it cost if it is on sale for $70 off? _____

⑭ **Extended Response** Mark lives at 241 East
Broadway. Sara lives at 90 East
Broadway. How far apart do they live? _____

Explain. _____

⑮ The Edinger family planned to drive
475 miles on the first day of their
vacation. They drove 90 miles
before lunch. How much
farther do they have
to drive today?

_____

Game Play the **Rummage Sale Game** in pairs or small groups.

LESSON 9.7 Three-Digit Subtraction

Key Ideas

When subtracting three-digit numbers, you might need to regroup more than once.

Check the ones column first.

$$\begin{array}{r} 676 \\ -\ 258 \\ \hline \end{array}$$

Regroup if needed. First try the tens place, and then the hundreds place.

$$\begin{array}{r} {}^{6}6\overset{1}{\cancel{7}}\cancel{6} \\ -\ 258 \\ \hline 18 \end{array}$$ detail ➝ $6\ \overset{6}{\cancel{7}}\ 16$

Subtract.

$$\begin{array}{r} 676 \\ -\ 258 \\ \hline \end{array}$$ $$\begin{array}{r} 676\ \text{letters} \\ -\ 258\ \text{letters} \\ \hline \text{\_\_\_\_ letters} \end{array}$$

Subtract. The first exercise has been started for you.

1.
$$\begin{array}{r} \overset{6}{4\cancel{7}13} \\ -218 \\ \hline \end{array}$$

2.
$$\begin{array}{r} 675 \\ -384 \\ \hline \end{array}$$

3.
$$\begin{array}{r} 582 \\ -272 \\ \hline \end{array}$$

4.
$$\begin{array}{r} 504 \\ -217 \\ \hline \end{array}$$

5.
$$\begin{array}{r} 608 \\ -319 \\ \hline \end{array}$$

6.
$$\begin{array}{r} 705 \\ -204 \\ \hline \end{array}$$

7.
$$\begin{array}{r} 600 \\ -357 \\ \hline \end{array}$$

8.
$$\begin{array}{r} 200 \\ -169 \\ \hline \end{array}$$

9.
$$\begin{array}{r} 410 \\ -328 \\ \hline \end{array}$$

10.
$$\begin{array}{r} 892 \\ -538 \\ \hline \end{array}$$

11.
$$\begin{array}{r} 479 \\ -265 \\ \hline \end{array}$$

12.
$$\begin{array}{r} 503 \\ -247 \\ \hline \end{array}$$

13. Explain a way to check your answers.

_____

_____

14. **Extended Response** Explain how you know when to regroup.

_____

_____

Game Play the **Checkbook Game.**

LESSON 9.8 — Applications of Three-Digit Addition and Subtraction

Key Ideas

Addition and subtraction can help you understand practical situations.

Use the table to answer the questions.

Every year Big City has a 15-mile race.

Total Number	Male	Female	Age			
			12–17	18–39	40–65	66+
Start 745	347	398	128	439	165	13
Finish 525	239	286	88	335	99	3

❶ How many people finished the race? _____

❷ How many did not finish? _____

❸ How many people younger than 40 finished? _____

❹ How many 18-year-old boys finished the race? _____

❺ How many people older than 65 finished? _____

❻ How many people older than 65 did not finish? _____

How much will these cost this month?

7 Computer _____

8 Printer _____

9 Stereo _____

10 Television _____

11 A computer and a printer _____

12 **Extended Response** How much money would you save by buying the computer and printer now instead of next month? Show two different ways to get your answer. _____

_____

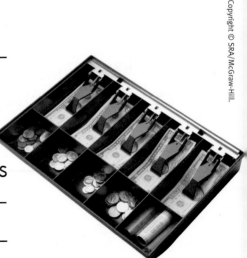

Name _____   Date _____

Listen to the problem.

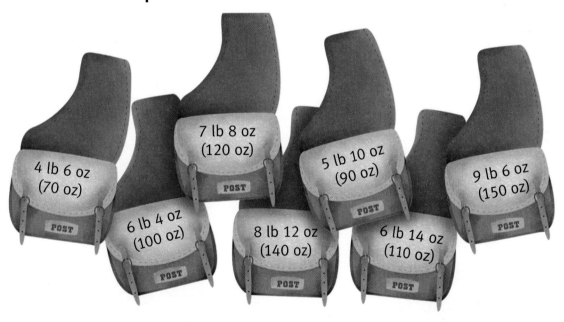

4 lb 6 oz (70 oz)

7 lb 8 oz (120 oz)

5 lb 10 oz (90 oz)

9 lb 6 oz (150 oz)

6 lb 4 oz (100 oz)

8 lb 12 oz (140 oz)

6 lb 14 oz (110 oz)

Lucy is using guess, check, and revise to solve the problem.

```
  70
 120    110
  90    140
+150   +100

 430
```

Cory is making a physical model to solve the problem.

10 × 15
(150 oz)

10 × 9
(90 oz)

10 × 12
(120 oz)

10 × 14
(140 oz)

10 × 7
(70 oz)

10 × 10
(100 oz)

10 × 11
(110 oz)

Show how you will solve the problem.

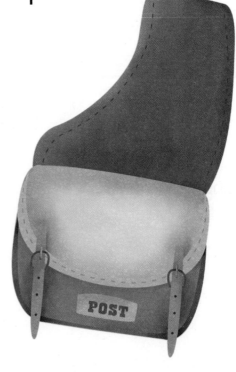

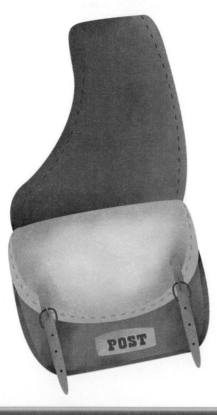

Cumulative Review

Name _____ Date _____

Measuring Length—Centimeters Lesson 4.2

Answer.

1 **Extended Response** Explain how to find the perimeter of a square.

_____

Place Value through 10,000 Lesson 9.1

Mark a point on the number line where you think each number belongs, and label it.

0   10

2 6, 2, 3, 5

3 19, 80, 91, 43

0   100

Fractions of Numbers Lesson 7.8

Write the number.

4 $\frac{2}{2}$ of 60 = _____   **6** $\frac{1}{4}$ of 60 = _____

5 $\frac{1}{2}$ of 100 = _____   **7** $\frac{4}{4}$ of 60 = _____

Regrouping for Addition Lesson 5.3

Write the number.

8  = _____   **9**  = _____

Cumulative Review

Comparing Numbers through 1,000 **Lesson 9.3**

Draw the correct sign.

10 118 ◯ 120

11 780 ◯ 840

12 200 + 20 ◯ 200 + 10

13 650 − 30 ◯ 650 − 40

14 550 ◯ 660

15 200 − 20 ◯ 200 − 10

Practicing Two-Digit Subtraction **Lesson 6.6**

Solve these problems.

16 Maria had $81. She spent $43. Now she has _____.

17 Jill has $34. She needs $67. She needs _____ more.

18 Alonzo has 3 dimes and 3 pennies. An eraser costs
35 cents. How much more money does Alonzo need
to buy an eraser? _____

Obtuse, Acute, and Right Angles **Lesson 8.5**

Identify the angles.

19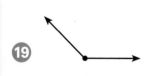

20

21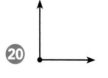

22

_____   _____   _____   _____

Adding Multiples of 10 **Lesson 9.4**

Write the number.

23 Seven hundred thirty-seven _____

24 Eight hundred ninety-three _____

Name _____ Date _____

 **LESSON 9.9** **Adding and Subtracting Money**

Key Ideas

$1 = 100 cents $1.25 = 125 cents

Write how much.

1 _____ cents = $2.00   **4** _____ cents = $0.96

2 _____ cents = $10.00   **5** _____ cents = $2.43

3 _____ cents = $1.50   **6** _____ cents = $5.00

7 **Extended Response** How many cents do the horse and the hat cost altogether? _____ cents

$$\begin{array}{r} 650 \\ + 75 \\ \hline \end{array}$$

Explain how you wrote the answer.

_____

_____

_____

8 Leslie has $10. Suppose she buys the horse and the hat with a $10 bill. How much change will she get? _____ cents

Copyright © SRA/McGraw-Hill.

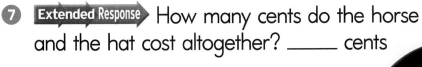

Solve the following problems.

9 John earns $2.75 an hour for working in the garden. How much does he earn in 2 hours? _____

10 Jason had $5.50. He spent some of the money. Now he has $2.75. How much did he spend? _____

11 A large container of juice costs $2.35. A smaller container costs $1.45. Which is the better buy? _____

12 Regular admission to the ball game is $4.75. Admission for children is $2.30. How much will it cost a father and his young daughter to go to the ball game together? _____

13 April bought a book for $3.75. She gave the cashier $5.00. How much change should she get? _____

14 Mark bought a puzzle for $3.76 and a pen for $1.25. How much did he pay for both? _____

LESSON 9.10

Shortcuts for Three-Digit Addition and Subtraction

Key Ideas

There are shortcuts for three-digit addition and subtraction problems.

"299 + 399 = ?
That's 2 less than 700."

"299 + 399 = ?
That's the same as 300 + 400 − 2."

Use shortcuts if you can.

Add.

1. $300 + 400 =$ _____

2. $300 + 399 =$ _____

3. $299 + 400 =$ _____

4. $299 + 399 =$ _____

5. $200 + 500 =$ _____

6. $201 + 499 =$ _____

7. $499 + 201 =$ _____

8. $499 + 202 =$ _____

Subtract.

9. $400 - 100 =$ _____

10. $400 - 99 =$ _____

11. $399 - 99 =$ _____

12. $401 - 99 =$ _____

13. $375 - 100 =$ _____

14. $374 - 99 =$ _____

15. $373 - 98 =$ _____

16. $372 - 97 =$ _____

eTextbook This lesson is available in the **eTextbook**.

Game

Addition and Strategies Practice

Make 1,000 Game

Players: Two or more

Materials:

- Pencil and paper
- **Number Cubes** (two 0–5 and two 5–10)

HOW TO PLAY

❶ Each player chooses and writes down any number between 250 and 750. This is the starting number.

❷ Player One rolls all four cubes. All players use the numbers rolled to make a one-, two-, or three-digit number. If a 10 is rolled, that cube is rolled again.

❸ All players add the number they made to their starting number. Everyone must do this at the same time without sharing information, and players are not permitted to change numbers after they have been chosen.

❹ The goal is to get as close to 1,000 as possible without going over 1,000.

❺ The winner of the round is the person who gets closest to 1,000 without going over. That person gets to roll the cubes in the next round of play.

ⓔ Games This game is available as an *eGame.*

LESSON 9.11 Round to 10 and 100 with Applications

Key Ideas

Rounding numbers is sometimes useful.

Sometimes we decide to round up.
Sometimes we decide to round down.
Sometimes we decide not to round because
we need an exact answer.

Whether we round up or down or not at all depends on
the situation. It depends on how we will use the numbers.

Mario has $700. He wants to
know if he has enough money
to buy the computer and printer.
He rounds up both prices to the
nearest hundred.

479 rounds to 500
134 rounds to 200

1 500 + 200 = _____

Brown's Computer Store	
Computer	$479
Printer	$134
Computer table	$287
DVD case	$45
Super software package	$297

The computer and printer would cost less than $700.
Mario has enough money.

Mrs. Brown owns the computer store. She knows
that she will make a profit if she can sell the
computer and printer for at least $450. To find
out, she rounds down to the nearest hundred.

479 rounds to 400
134 rounds to 100

2 $400 + 100 =$ _____
$500 is more than $450. Mrs. Brown will make a profit.

Susan wants to know the approximate cost of the computer and printer. To find out, she rounds both prices to the nearest ten.

479 is closer to 480 than to 470. She rounds up to 480.
134 is closer to 130 than to 140. She rounds down to 130.

Susan adds the rounded numbers.

3 $480 + 130 =$ _____
The computer and printer cost about $_____.

Susan decides to buy the computer and printer, so she needs the exact price.

4 $479 + 134 =$ _____
The computer and printer will cost exactly $_____.

Round each number to the nearest hundred.
If the number is halfway between, round
up. The first exercise is done for you.

5 745 700

6 639 _____

7 210 _____

8 453 _____

9 650 _____

10 180 _____

Round each number to the nearest ten. The first exercise is
done for you. If the number is halfway between, round up.

11 379 380

12 524 _____

13 671 _____

14 454 _____

15 275 _____

16 823 _____

LESSON 9.12 Approximating Answers

Key Ideas

The answer to some problems can be approximated by using simpler numbers.

$297 + 305 =$
297 is almost 300, and 305 is just a little more than 300.

We know $300 + 300 = 600$.
So the sum of $297 + 305$ is about 600.

Ring the letter of the correct answer.
The first problem has been done for you.

1
$$\begin{array}{r} 98 \\ + 103 \end{array}$$
a. 101
b. 201 (circled)
c. 301

2
$$\begin{array}{r} 407 \\ + 398 \end{array}$$
a. 605
b. 705
c. 805

3
$$\begin{array}{r} 489 \\ + 312 \end{array}$$
a. 599
b. 701
c. 801

4
$$\begin{array}{r} 482 \\ + 312 \end{array}$$
a. 694
b. 704
c. 794

5
$$\begin{array}{r} 306 \\ + 247 \end{array}$$
a. 453
b. 503
c. 553

6
$$\begin{array}{r} 694 \\ + 156 \end{array}$$
a. 750
b. 800
c. 850

e Textbook This lesson is available in the *eTextbook*.

Solve these problems.

7 **Extended Response** Max has $9.00. He will buy flowers for $7.98. How much change should he get? Ring the letter of the correct answer.

a. $1.02   **b.** $8.02   **c.** $10.02

Explain how you solved this problem. _____

_____

8 **Extended Response** Karen earned $3.25 yesterday. She earned $4.50 today. How much did she earn in the two days? Ring the letter of the correct answer.

a. $6.05   **b.** $7.05   **c.** $7.75

Explain how you solved this problem. _____

_____

89¢ a bunch

$1.79 a pound

75¢ a pound

Jim has $5. Does he have enough money to buy

9 2 pounds of strawberries? _____   **11** 3 pounds of pears? _____

10 3 bunches of grapes? _____   **12** 4 pounds of strawberries? _____

Name _____ Date _____

Listen to the problem.

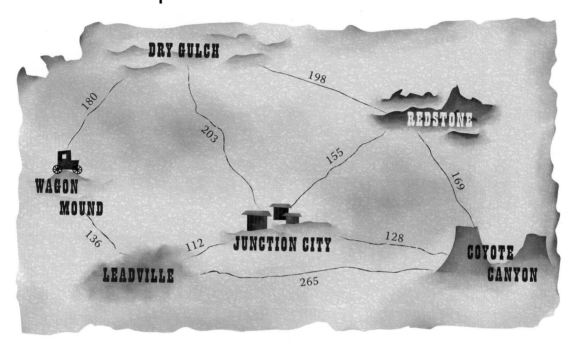

Show your route. Show how you know it is less than 1,000 miles.

Start and end at Silver Falls.

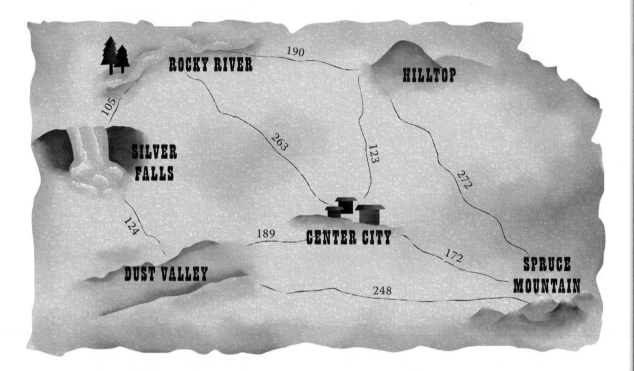

Show your route. Show how you know it is less than 1,000 miles.

Cumulative Review

Name _____ Date _____

Measuring Length—Centimeters Lesson 4.2

Solve. Use your centimeter ruler to measure.
Write your answers on the lines.

❶ Triangle

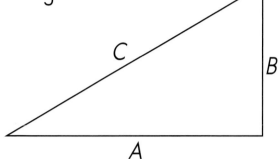

Side	Centimeter
A	_____
B	_____
C	_____

Perimeter _____

Fractions of a Set Lesson 7.5

Write the fraction shown.

❷ ⬤⬤⬤〇〇〇 _____

❸ ⬠⬠⬠⬠⬠⬠ _____

❹ ◼◼◼◼◼☐ _____

❺ ▲△ _____

Cumulative Review

Solve these problems.

6 Jamal's digital camera can hold 90 pictures.
He has already taken 39 pictures.
How many more pictures can he take? _____

7 Keisha is allowed to use her cell phone for
80 minutes each week. She has already used
67 minutes. Can Keisha make a 15-minute
phone call? _____ Explain. _____

_____

Applications of Three-Digit Addition and Subtraction Lesson 9.8

How much will these cost today?

8 tractor _____

9 tilling machine _____

10 tractor cart _____

11 a tractor and a tractor cart _____

Key Ideas Review

Name _____ Date _____

In this chapter you learned about three-digit addition and subtraction. You learned to compare numbers through 1,000. You also learned about adding and subtracting money.

Write your answers.

❶ In the number 6,254 there are

   a. how many thousands? ____   **b.** how many hundreds? ____

   c. how many tens? ____   **d.** how many ones? ____

❷ Write the number that has 3 tens,
   5 hundreds, 8 ones, and 1 thousand. _____

❸ Write a number in the circle to make the inequality true.

   $670 <$ ◯

Solve each problem.

❹   276
    + 395

❺   606
    − 258

Write your answers.

The sunglasses cost $3.25.   The paint set costs $4.50.

6 How many cents do the sunglasses
   and paint set cost together?   _____ cents

7 Stacy has $10. If she buys the
   sunglasses and the paint set,
   how much change will she get?   _____ cents

8 Explain a shortcut for solving 299 + 399.

   _____

9 Tim rounded 730 to 800.
   Why would 700 probably have been a better answer?

   _____

10 How can you estimate the answer to
   297 + 305 without actually adding the numbers?

   _____

   _____

Name _____ Date _____

Lesson 9.11 **Round** to nearest 100.

① 480 _____   ② 390 _____   ③ 120 _____

④ 70 _____   ⑤ 840 _____   ⑥ 540 _____

⑦ **Extended Response** Is 450 closer to 500 or to 400? Explain.

_____

Lesson 9.1 **Write** the numbers.

⑧ In 843
How many hundreds? _____

How many tens? _____

How many ones? _____

⑨ In 3,184
How many thousands? _____

How many hundreds? _____

How many tens? _____

How many ones? _____

Lesson 9.12 **Ring** the best approximate answer.

⑩ 127	⑪ 407	⑫ 512
+ 43	+ 398	+ 389
175 150 200	700 800 900	915 815 715

Lesson 9.3 **Write** the words *greater than*, *less than*, or *equal to*.

⑬ Five hundred fifty is _____ four hundred sixty.

⑭ Three hundred is _____ five hundred.

Lesson 9.9 **Write** how much.

⑮  _____ cents

⑯ 84 cents = _____

⑰ 206 cents = _____

⑱ $1.21 = _____ cents

Lessons 9.5–9.8 **Add** or subtract. Watch the signs.

⑲ 243
+ 621

⑳ 447
+ 293

㉑ 161
+ 820

㉒ 724
+ 893

㉓ 12
− 6

㉔ 56
− 4

㉕ 42
− 7

㉖ 420
− 70

Practice Test

Name _____ Date _____

Write the missing numbers.

1 701, 702, _____, _____, _____, 706

2 257, 258, _____, _____, _____, 262

3 1,123, _____, 1,125, _____, _____, 1,128

4 1,609, _____, 1,611, _____, _____, 1,614

Draw >, <, or =.

5 557 ⃝ 487

7 630 + 30 ⃝ 630 + 40

6 161 ⃝ 159

8 428 + 12 ⃝ 430 + 10

Write how much.

9 536 cents = $_____

10 $2.17 = _____ ¢

11 337 cents = $_____

12 $6.09 = _____ ¢

Ring the letter of the number words that match the numbers.

13 462
   a. four hundred sixty-two
   b. four hundred twenty-six
   c. four hundred sixteen
   d. four hundred sixty

14 291
   a. one hundred ninety-two
   b. one hundred twenty-nine
   c. two hundred ninety-one
   d. two hundred nineteen

Add.

15 312 + 190 =
   a. 122   **b.** 331   **c.** 502   **d.** 522

16 724 + 264 =
   a. 460   **b.** 848   **c.** 888   **d.** 988

17 521 + 309 =
   a. 910   **b.** 830   **c.** 560   **d.** 212

Subtract.

18 880 − 90 =
   a. 690   **b.** 790   **c.** 800   **d.** 970

Name _____ Date _____

Subtract.

⑲ 628 − 293 =

    a. 335   **b.** 435
    c. 475   **d.** 921

⑳ 900 − 42 =

    a. 742   **b.** 758
    c. 842   **d.** 858

Add or subtract to solve.

㉑ Cathy's Caterers made 500 chicken wings for a dinner. They served 448 of the wings and gave away 1 dozen wings. How many wings were left?

    a. 52   **b.** 51   **c.** 50   **d.** 40

㉒ Rita practiced her violin for 46 minutes on Monday and 29 minutes on Tuesday. How many minutes did she practice on Monday and Tuesday altogether?

    a. 75   **b.** 23   **c.** 20   **d.** 17

㉓ Lance is 158 centimeters tall. Lionel is 190 centimeters tall. How many centimeters shorter is Lance than Lionel?

    a. 22   **b.** 32   **c.** 48   **d.** 348

Ring the time.

㉔   **a.** half past 1
    b. quarter to 2
    c. quarter to 1
    d. quarter after 2

Ring the time.

25

a. quarter to 11
b. quarter to 12
c. quarter after 11
d. quarter after 12

26

a. 3:15
b. 3:45
c. 9:30
d. 9:15

Use the table to solve.

  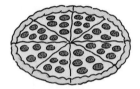

Sandwich	$4.75
Pizza, slice	$2.50
Pizza, whole	$9.00
Milk	100 cents

27 Paul spent about 600¢ for lunch. List two ways he could have spent his money on his lunch. Then write how much he would have spent for each possible lunch.

_____

_____

_____

28 Vanessa and three of her friends had $20 altogether to spend for lunch. Write what they could have bought. How much did they spend? How much was left?

_____

_____

_____

Thinking Story

Loretta the Letter Carrier Chooses Sides

Ring the letters with even-numbered addresses.

Complete the pattern.

1 2 4 6 8 10 12 ____ ____ ____

2 17 15 13 11 9 7 ____ ____ ____

3 5 5 7 7 9 9 ____ ____ ____

4 10 8 9 8 6 7 ____ ____ ____

5 1 2 4 7 11 16 ____ ____ ____

Real Math • Chapter 9

CHAPTER 10 Measurement

In This Chapter You Will Learn

- to read a thermometer and a map.
- about weight and capacity.
- to tell time.

Name _____ Date _____

Listen to the problem.

My plan for the largest garden:

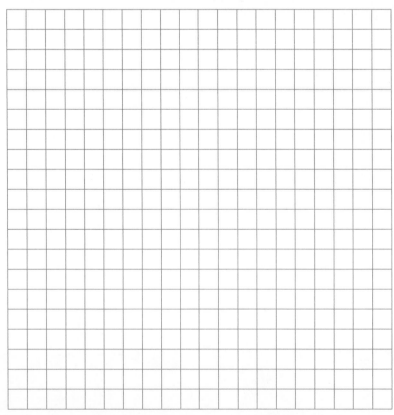

LESSON 10.1

Reading a Thermometer

Key Ideas

A thermometer measures temperature. Only even-numbered degrees are marked on most thermometers.

Measure the temperature to complete the activity.

Cup 1   Cup 2   Cup 3

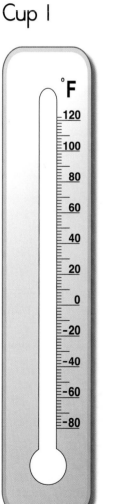

_____°   _____°   _____°   _____°

                    Prediction   Actual

e Textbook This lesson is available in the *eTextbook*.

Write or show the temperature.

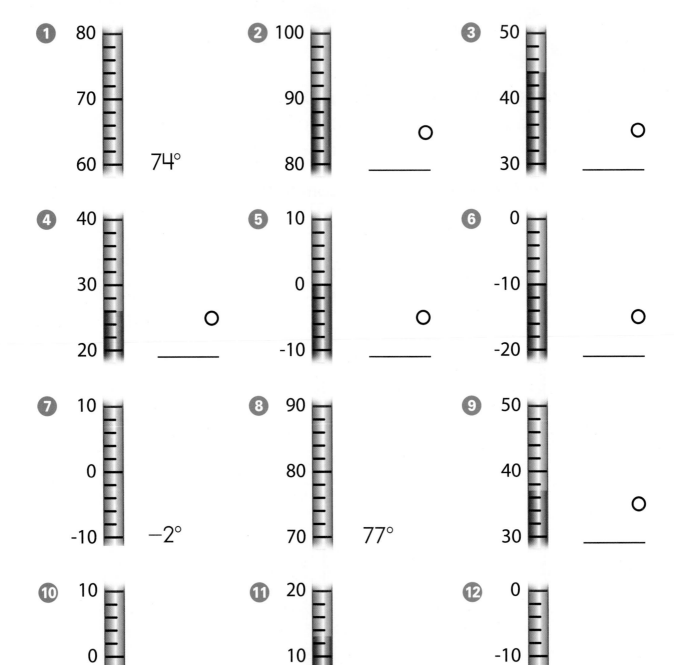

1. 80 70 60 74°

2. 100 90 80 ____ ○

3. 50 40 30 ____ ○

4. 40 30 20 ____ ○

5. 10 0 -10 ____ ○

6. 0 -10 -20 ____ ○

7. 10 0 -10 −2°

8. 90 80 70 77°

9. 50 40 30 ____ ○

10. 10 0 -10 5°

11. 20 10 0 ____ ○

12. 0 -10 -20 ____ ○

Copyright © SRA/McGraw-Hill.

LESSON 10.2 Reading a Map

Key Ideas

Maps help us get where we want to go.

Study the map.

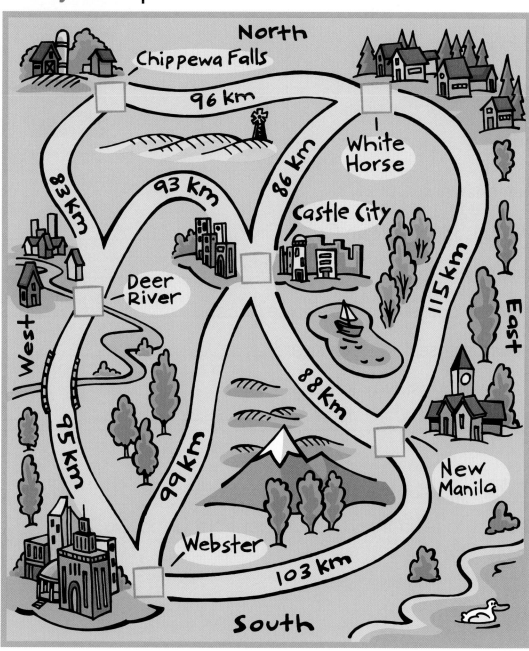

Complete the table. Use the map on page 377 to find the shortest distances.

	Towns	Shortest Distance (kilometers)
1	Chippewa Falls and New Manila	
2	New Manila and Deer River	
3	White Horse and Webster	
4	Chippewa Falls and Webster	

Write your answers.

5 Which town is farthest from Castle City? _____

6 Which town is farthest from Webster? _____

7 Which town is closest to White Horse? _____

8 Which town is closest to Deer River? _____

9 Which is the shortest way to get to Deer River from White Horse? _____

 **Journal**

Write directions to tell a friend how to get from Chippewa Falls to Castle City.

LESSON 10.3 Measuring Perimeter

Key Ideas

The perimeter is the measure of the distance on a path around a figure.

Find the perimeter.

1

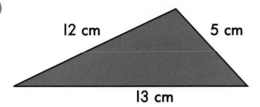

Perimeter = _____ cm

2

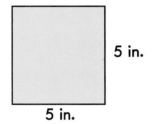

Perimeter = _____ in.

3

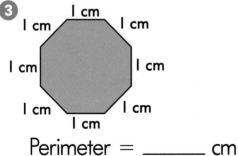

Perimeter = _____ cm

4

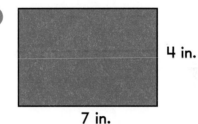

Perimeter = _____ in.

5

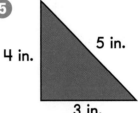

Perimeter = _____ in.

6

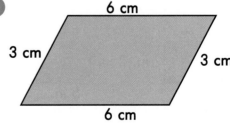

Perimeter = _____ cm

eTextbook This lesson is available in the *eTextbook.*

Measure objects in your classroom. Fill in the table.

	Object	Perimeter in Centimeters	Perimeter in Inches
7			
8			
9			

What is the perimeter of the combined figures?

10

3 in. /\ 3 in. 3 in. /\ 3 in.
3 in.   3 in.

_____ Perimeter of both together

_____ Perimeter

11

4 in.
4 in. ☐ 4 in.   4 in. ☐ 4 in.
4 in.   4 in.

_____ Perimeter of both together

_____ Perimeter

12

5 in.   5 in.
2 in. ☐ 2 in. 2 in. ☐ 2 in.
5 in.   5 in.

_____ Perimeter of both together

_____ Perimeter

LESSON 10.4 Kilograms and Grams

Key Ideas

The kilogram and the gram are units of weight.
There are 1,000 grams in 1 kilogram.

How many?

1 1 kilogram = _____ grams

2 2 kg = _____ g

3 3 kg = _____ g

4 4 kg = _____ g

5 5 kg = _____ g

6 6 kg = _____ g

ⓔ Textbook This lesson is available in the *eTextbook*.

How much does it weigh?

Record your answers on the table.

Object	Estimated Weight	Measured Weight	Difference

LESSON 10.5 Pounds and Ounces

Key Ideas

Pounds and ounces are units of weight.
There are 16 ounces in 1 pound.

about 1 pound

...between 1 and 2 ounces...

How many?

① How many ounces in 1 pound? _____

② How many ounces in 2 pounds? _____

③ How many ounces in 4 pounds? _____

④ How many ounces in 8 pounds? _____

⑤ How many ounces in 16 pounds? _____

Write your answers.

6 Which box probably weighs the most?

_____

7 Which box probably weighs the least?

_____

How much does it weigh? Record your answers on the table.

Object	Estimated Weight	Measured Weight	Difference

Name _____ Date _____

Listen to the problem.

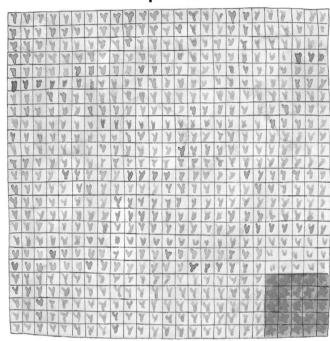

How many bags of fertilizer are needed for the rest of the garden?

Linda is making a physical model to solve the problem.

Carlos is making a diagram to solve the problem.

 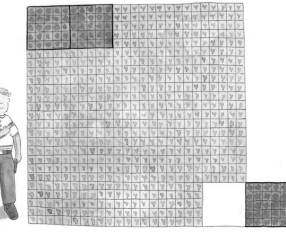

Solve the problem.

They should buy _____ bags of fertilizer.

Show how you know.

Cumulative Review

Name _____ Date _____

Pictographs Lesson 4.9

Use the pictograph to answer the questions below.

Each gas can stands for 10 gallons of gas.

Ava Island to Murphy	🛢️ 🛢️ 🛢️ 🛢️ 🛢️ 🛢️ 🛢️ 🛢️ 🛢️ 🛢️
Murphy to Brushy Fork	🛢️ 🛢️ 🛢️ 🛢️ 🛢️ 🛢️
Brushy Fork to Flint	🛢️ 🛢️ 🛢️

1 According to the pictograph, which trip takes the least gas? _____

2 How many gallons are used on a trip from Murphy to Brushy Fork? _____

Applications of Addition and Subtraction Lesson 3.6
Solve.

3 Shelly can walk from her house to Pat's house in 4 minutes. About how many minutes will it take her to walk back? _____

_____

4 About how many minutes will it take her to walk to Pat's house and back? _____

_____

Kilograms and Grams Lesson 10.4
How many?

5 1 kilogram = _____ g **6** 5 kg = _____ g

Cumulative Review

Use the map to complete the table. Then use the table to answer the questions.

Nita marked towns on a map. She makes deliveries to the plant stores in these towns.

	Towns	Shortest Distance (kilometers)
7	Dover and Hiteston	
8	Newsail and Langston	
9	Dover and Langston	
10	Wayne and Dover	

11 What town is farthest from Dover? _____

12 What town is closest to Marshal? _____

13 What is the shortest way to get to Newsail from Wayne? _____

Grouping by Tens Lesson 5.9
Solve.

14 $20 + 40 + 60 + 80 + 100 = $ _____

15 $10 + 9 + 8 + 2 + 1 + 10 = $ _____

16 $7 + 3 + 1 + 9 + 5 + 5 + 3 = $ _____

17 $4 + 6 + 7 + 3 + 5 + 5 + 9 + 1 = $ _____

LESSON 10.6

Measuring Capacity (Customary Units)

Key Ideas

Capacity is a measurement of the amount a container can hold.

You can express capacity in different ways.

2 pints
= 1 quart

8 fluid ounces = 1 cup

16 fluid ounces = 1 pint

4 quarts
= 1 gallon

How many?

1 How many fluid ounces in 1 quart? _____

2 How many cups in 1 pint? _____

3 How many pints in 1 gallon? _____

4 How many fluid ounces in 1 gallon? _____

5 How many fluid ounces in 2 gallons? _____

6 How many fluid ounces in 3 gallons? _____

7 How many fluid ounces in 4 quarts? _____

8 How many pints in 4 quarts? _____

ⓔ Textbook This lesson is available in the *eTextbook*.

Number the pictures in order from smallest capacity to largest capacity.

1 pint
2 cups or
16 fluid ounces

1 quart
2 pints

1 cup
8 fluid ounces

1 fluid ounce

1 gallon
4 quarts

_____   _____   _____   _____   _____

Match equal measures.

2 pints

4 quarts

3 gallons

8 pints

12 quarts

32 fluid ounces

LESSON 10.7 Measuring Capacity (Metric Units)

Key Ideas

Milliliters and liters are also units of capacity.
There are 1,000 milliliters in 1 liter.

How many?

1 _____ milliliters = 2 liters

2 4 liters = _____ milliliters

3 5,000 milliliters = _____ liters

4 $\frac{1}{2}$ liter = _____ milliliters

5 3 liters = _____ milliliters

6 _____ liters = 6,000 milliliters

Ring the better estimate for the capacity of each object.

7

about 300 milliliters

about 3 liters

8

about 2 liters

about 200 liters

9

about 200 milliliters

about 2 liters

10

about 90 milliliters

about 900 milliliters

LESSON 10.8 **Units of Time**

Key Ideas

There are seven days in one week.

Fill in the missing days.

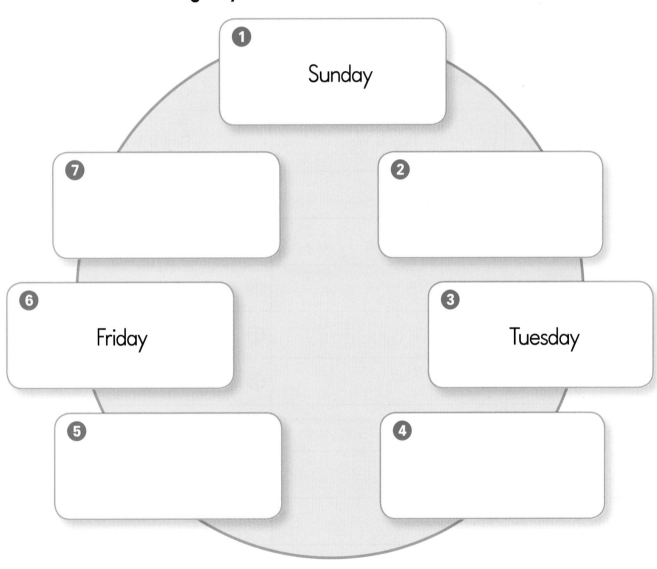

1. Sunday
2.
3. Tuesday
4.
5.
6. Friday
7.

What day is it today?

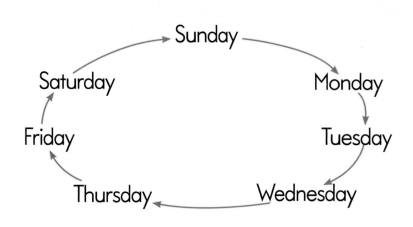

Use today to complete the table. Write the day it was and the day it will be.

Number of Days	Day It Was	Day It Will Be
1	**8**	**9**
3	**10**	**11**
6	**12**	**13**
7	**14**	**15**
8	**16**	**17**
14	**18**	**19**
21	**20**	**21**
69	**22**	**23**
70	**24**	**25**
71	**26**	**27**

Name _____ Date _____

Telling Time to the Minute

Key Ideas

Each line on the clock represents 1 minute.
There are 60 minutes in 1 hour.

What time is it?

1

_____ : _____

5

_____ : _____

2

_____ : _____

6

_____ : _____

3

_____ : _____

7

_____ : _____

4

_____ : _____

8

_____ : _____

Show the time on each clock.

9 3:27

11 Four minutes past eight.

10 4:51

12 Three minutes before seven

LESSON 10.10 **Elapsed Time**

Key Ideas

You can figure out what time it was in the past or what time it will be in the future.

Times between midnight and noon are A.M.
Times between noon and midnight are P.M.

The time is now 4:00 P.M.
Complete the table.

Number of Hours	Time It Was	Time It Will Be
1	3:00 P.M.	5:00 P.M.
2	❶	❷
4	❸	❹
8	❺	❻
12	❼	❽
24	❾	❿
36	⓫	⓬
48	⓭	⓮
72	⓯	⓰
73	⓱	⓲

ⓔTextbook This lesson is available in the *eTextbook*.

19 Mrs. Fontana gave her class an assignment on Friday and said it must be finished in eight days. What day of the week will that be?

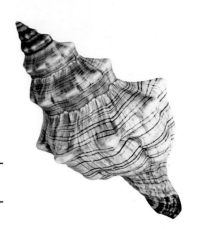

_____

20 What's wrong with Mrs. Fontana's assignment?

_____

_____

How many seconds does it take?

First estimate. Fill in the table below.

Task	Estimate (in seconds)	Measure (in seconds)
21 Count to 50.		
22 Count to 25.		
23 Say the sentence at the bottom of the page clearly.		
24 Say the sentence at the bottom of this page clearly twice.		

She sells seashells by the seashore.

Name _____ Date _____

Listen to the problem.

½ CUP

Ring the correct choice.

Joey **will** **will not**
be finished in time
for lunch at 12:00.

Show how you know.

Ring the correct choice.

Lane **will will not** be finished in time for lunch at 1:00.

This is how I know.

Cumulative Review

Name _____ Date _____

Writing Fractions Lesson 7.2

Write the fraction of each museum pass that is shaded.

1 [shaded fraction diagram] _____   **2** [shaded fraction diagram] _____   **3** [shaded fraction diagram] _____

Subtraction Facts and the Addition Table Lesson 3.2

Answer the questions.

$4 $2 $3 $6 $14

4 A person wants to buy the tricycle and the horse. If she gives the store a $20 bill, how much money will she get back? _____

5 How much would one toy plane cost? _____

Place Value through 10,000 Lesson 9.1

Mark a point on the number line where you think each number belongs, and label it.

6 306, 905, 400, 830, 100

0 ◄─────────────────────────────────────► 1,000

7 100, 500, 360, 800, 200

0 ◄─────────────────────────────────────► 1,000

Cumulative Review

Units of Time Lesson 10.8

Fill in the missing days.

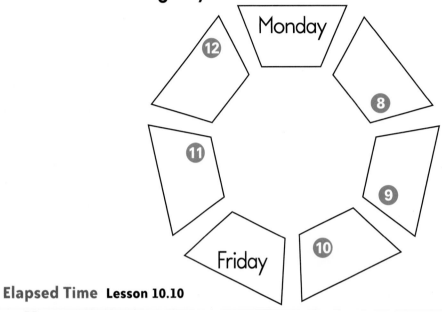

Elapsed Time Lesson 10.10

Fill in the table.

Hannah's landscaping company has ten different sprinklers. Each sprinkler runs for a certain number of hours. Finish the table.

Sprinkler	Total Hours	Start Time	Finish Time
Sprinkler 1	2 hours	8:00 A.M.	⑬
Sprinkler 2	2 hours	11:00 A.M.	⑭
Sprinkler 3	6 hours	10:00 A.M.	⑮
Sprinkler 4	7 hours	9:00 A.M.	⑯
Sprinkler 5	10 hours	1:00 P.M.	⑰
Sprinkler 6	15 hours	9:00 A.M.	⑱
Sprinkler 7	18 hours	2:00 P.M.	⑲
Sprinkler 8	24 hours	5:00 P.M.	⑳
Sprinkler 9	30 hours	3:00 P.M.	㉑
Sprinkler 10	48 hours	12:00 P.M.	㉒

Name _____ Date _____

In this chapter you learned about measurement.
You learned about measuring temperature, distance, weight, capacity, and time. You learned many different units of measurement.

Draw a line from each unit of measurement to the quantity it measures.

1 degrees   weight

2 liters   time

3 days   temperature

4 pounds   capacity

Write your answer.

5 Martina has two bags of grapes.
The first bag weighs 500 grams.
The second bag weighs 1 kilogram.
Which bag do you think has more grapes in it? _____

Why? _____

Key Ideas Review

Ring the best estimate for each.

6

more than 1 pound

less than 1 pound

8

about 100 grams

about 100 kilograms

7

about 1 gallon

about 1 pint

9

about 3 milliliters

about 300 milliliters

Follow the directions.

10 Isaac went to his friend's house at 2:15. His mom will pick him up $2\frac{1}{2}$ hours later. Draw the hands on the clock to show the time when Isaac's mom will pick him up.

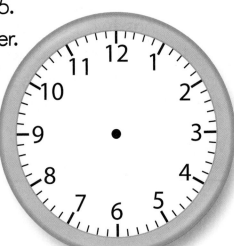

Name _____ Date _____

Lesson 10.5 **Ring** the best choice.

① How many ounces in a pound?

   1 ounce   16 ounces

② How many pounds in 48 ounces?

   4 pounds   3 pounds

③ How many ounces in 2 pounds?

   32 ounces   42 ounces

④ How much does an eraser weigh?

   More than 1 pound
   Less than 1 pound

Lesson 10.1 **Count** by twos. Fill in the blanks.

⑤ 26, _____, _____, _____, _____, 36

⑥ 90, _____, _____, _____, _____, 100

⑦ −20, −18, −16, _____, _____, −10

⑧ −30, −28, _____, _____, _____, −20

Lesson 10.9 **What** time is it?

⑨

⑩

⑪

____ : ____   ____ : ____   ____ : ____

Lesson 10.3 **Find** the perimeter.

12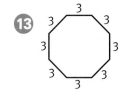

16   7

19

Perimeter = _____

13

3 3 3
3   3
3   3
3 3 3

Perimeter = _____

14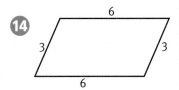

6
3   3
6

Perimeter = _____

15 You must dig a square hole with a perimeter of at least 32 cm. Draw the hole on the picture. What is the length of each side? _____

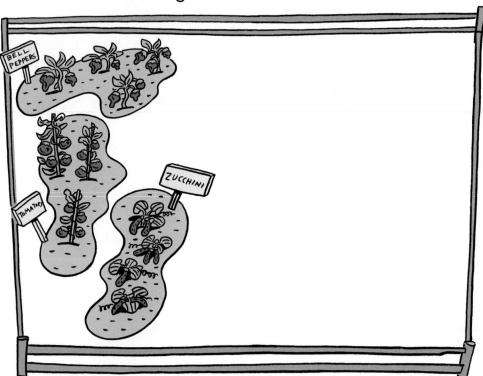

Lessons 10.6–10.7 **How many?**

16 _____ milliliters = 2 liters

17 4 liters = _____ milliliters

18 5,000 milliliters = _____ liters

19 _____ fluid ounces = 1 quart

20 _____ fluid ounces = 1 gallon

21 _____ cups = 1 pint

Practice Test

Name _____ Date _____

Write the temperature.

1 100 ⟋ 90 ⟋ 80 ___

2 10 ⟋ 0 ⟋ -10 ___

3 50 ⟋ 40 ⟋ 30 ___

4 10 ⟋ 0 ⟋ -10 ___

Find the perimeter.

5 Mr. Rojas put a border around the class bulletin board. The board is 60 inches long and 48 inches wide. How much border did it take?   _____ inches

6 Erika bought a border to go around her flower garden. The garden is 11 feet long and 4 feet wide. How many feet of border did she need?   _____ feet

Write the time.

7

8

9

___:___   ___:___   ___:___

Ring the letter of the answer that makes the comparison true.

10 4 kilograms = _____

    a. 1,000 grams   **b.** 100 grams
    c. 4,000 grams   **d.** 400 grams

11 4 quarts = _____

    a. 1 pint   **b.** 2 pints
    c. 1 gallon   **d.** 2 gallons

12 8 cups = _____

    a. 1 quart   **b.** 2 quarts
    c. 1 pint   **d.** 2 pints

13 9 liters = _____

    a. 9,000 milliliters   **b.** 900 milliliters
    c. 1,000 milliliters   **d.** 100 milliliters

Ring the letter of the best choice.

14 About how much do 4 quarters weigh?

    a. about 10 pounds   **b.** about 10 ounces
    c. about 1 pound   **d.** about 1 ounce

Name _____ Date _____

Ring the best answer.

⑮ Shelly is a hospital doctor. She goes to work at 7:00 P.M. She leaves work at 2:00 A.M. How many hours per day does Shelly work?

a. 5 hours   **b.** 6 hours
c. 7 hours   **d.** 9 hours

⑯ Ted visited his grandmother on Thursday. Van visited his grandmother 2 days earlier. On what day did Van visit his grandmother?

a. Saturday   **b.** Wednesday
c. Tuesday   **d.** Sunday

Ring the letter of the object that is closest to the length given.

⑰ 1 inch

a. shoe   **b.** paper clip
c. television   **d.** desk

⑱ 1 meter

a. 4-year-old child   **b.** tall woman
c. math book   **d.** train

Use the map to solve.

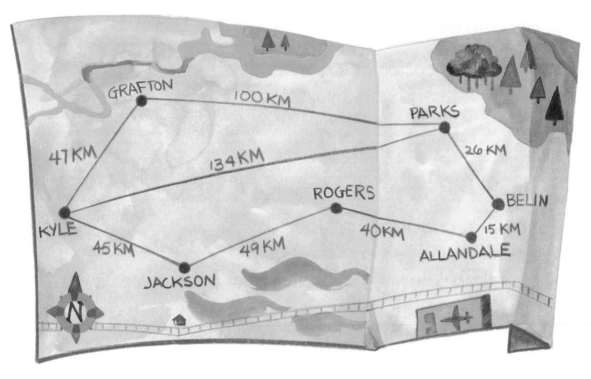

19 There is more than one way to get from Grafton to Rogers. What is the shortest way? Write the cities that you would go through and the total distance.

20 There is more than one way to get from Kyle to Belin. What is the shortest way? Write the cities you would go through and the total distance.

Thinking Story

Ferdie's Meterstick

Measure the height of the desk
in centimeters. Write the number. _____

Ring the items that are about 1 meter tall.

CHAPTER 11 Introducing Multiplication and Division

In This Chapter You Will Learn
- how to use repeated addition and arrays.
- what multiplication and division are.
- how to multiply and divide.

Name _____ Date _____

Listen to the problem.

The club spent _____ for 12 meals.

This is how I know:

LESSON 11.1

Skip Counting and Repeated Addition

Key Ideas

Skip counting is counting on in groups and leaving out the numbers in between.

This number line shows skip counting by 3s.

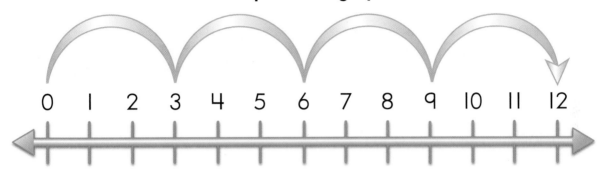

Skip count and fill in the missing numbers.

1 0 2 ____ ____ 8 10

2 0 5 10 ____ ____ 25

3 0 3 6 ____ ____ 15

4 0 4 ____ 12 ____ 20

5 If 3 tennis balls are in a can, use skip counting by threes to figure out how many tennis balls are in 5 cans.

eTextbook This lesson is available in the *eTextbook*.

Some problems can be solved with repeated addition.

Each slice of pizza has three sides. How many sides are there?

6   3 + 3 + 3 + 3 + 3 = _____

How many 3s did you add? _____

Each plate has four bagels. How many bagels are there?

7   4 + 4 + 4 + 4 = _____

How many 4s did you add? _____

Each plate has six pieces of sushi. How many pieces of sushi are there?

8   6 + 6 + 6 + 6 + 6 = _____

How many 6s did you add? _____

Writing + Math   **Journal**

How could you write the following skip-counting sequence as a repeated addition problem?

4, 8, 12, 16, 20

LESSON 11.2

Introduction to Multiplication

Key Ideas

There is a shorter way to write repeated addition problems

$2 + 2 + 2 = 6$

You can use **multiplication.**

$\times$ means "times."
2 is added 3 times
3 times 2 = 6
$3 \times 2 = 6$

number of times the   number being
number is repeated   added or repeated

It is also the same as multiplying 2 times 3:
$2 \times 3 = 6$.

Solve.

1 $8 + 8 + 8 + 8 + 8 + 8 + 8 =$ _____

How many times did you add 8? _____ times

So, $7 \times 8 =$ _____.

2 $3 + 3 + 3 + 3 + 3 =$ _____

How many times did you add 3? _____ times

_____ $\times$ _____ = _____

3 $5 + 5 + 5 =$ _____

How many times did you add 5? _____ times

_____ $\times$ _____ = _____

4 $3 =$ _____

How many times did you add 3? _____ times

_____ $\times$ _____ = _____

5 $1 + 1 + 1 =$ _____

How many times did you add 1? _____ times

_____ $\times$ _____ = _____

6 $3 + 3 + 3 + 3 =$ _____

How many times did you add 3? _____ times

_____ $\times$ _____ = _____

7 $4 + 4 + 4 =$ _____

How many times did you add 4? _____ times

_____ $\times$ _____ = _____

8 $2 + 2 + 2 + 2 + 2 + 2 =$ _____

How many times did you add 2? _____ times

_____ $\times$ _____ = _____

 **Journal**

What would be the matching repeated-addition expression for 3×8?

LESSON 11.3

Multiplication and Arrays

Key Ideas

Multiplication can be shown by arranging pictures in arrays.

$4 + 4 + 4 = 12$

So, $3 \times 4 = 12$.

Use these pictures to solve the problems.

1 _____ $\times 4 = 20$

3 _____ $\times 3 = 9$

2 _____ $\times 4 = 12$

4 _____ $\times 2 = 6$

e Textbook This lesson is available in the *eTextbook*.

Use these pictures to solve the problems.

⑤ $1 \times 4 =$ _____

⑥ $2 \times 4 =$ _____

⑦ $3 \times 4 =$ _____

⑧ $4 \times 4 =$ _____

⑨ $5 \times 4 =$ _____

⑩ $1 \times 3 =$ _____

⑪ $2 \times 3 =$ _____

⑫ $3 \times 3 =$ _____

⑬ $4 \times 3 =$ _____

⑭ $5 \times 3 =$ _____

⑮ **Extended Response** How many scones do you think this tray can hold? _____ Explain why you think there are that many.

_____

_____

LESSON 11.4 **Arrays and Area**

Key Ideas

You can understand area by studying arrays.

Area is the total number of square units within a figure.

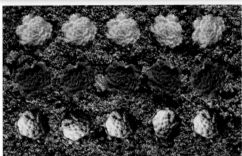

1 How many vegetable plants are in Marie's garden? _____

She built a fence around her garden.

2 What shape is Marie's garden? _____

3 How many vegetables tall is it? _____

How many vegetables wide is it? _____

Then Marie installed sprinklers to water all her plants. Marie's garden is now divided into square units.

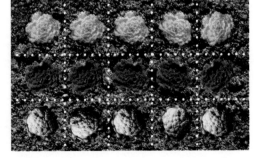

4 How many square units is Marie's garden? _____

5 What is the area of Marie's garden? _____ square units

6 Write a multiplication sentence to show how to find the area of Marie's garden.

_____

Use the squares to solve these problems.

10

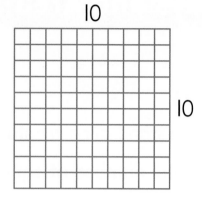

7 Area _____ square units

10 × 10 = _____

8 Area _____ square units

5 × 3 = _____

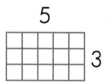

9 Area _____ square units

3 × 14 = _____

10 Extended Response Explain how you found the area.

_____

_____

_____

LESSON 11.5 Using the Multiplication Table

Key Ideas

Multiplication facts can be found on the Multiplication Table.

$3 \times 4 = 12$

✕	0	1	2	3	4	5
0	0	0	0	0	0	0
1	0	1	2	3	4	5
2	0	2	4	6	8	10
3	0	3	6	9	12	15
4	0	4	8	12	16	20
5	0	5	10	15	20	25

Use the Multiplication Table to solve.

1 $3 \times 5 =$ _____

2 $2 \times 3 =$ _____

3 $4 \times 4 =$ _____

4 $5 \times 5 =$ _____

5 _____ $\times 3 = 9$

6 _____ $\times 4 = 8$

7 _____ $\times 5 = 0$

8 _____ $\times 2 = 2$

9 $0 \times 3 =$ _____

⑩ **Extended Response** Carmen earns $4 each day for watching her younger brother after school. How much money does she earn for 3 days? _____

Explain how you know. _____

⑪ Heide, Hilla, and Heike each have $4. How many dollars do they have altogether? _____

⑫ Tomatoes are sold 4 to a box. How many tomatoes are in 5 boxes? _____

⑬ A square measures 5 centimeters on each side.

What is its perimeter? _____

What is its area? _____

⑭ Chester, Ms. Heather's cat, eats about 4 cans of cat food each week. About how many cans of cat food will Heather need for 3 weeks? _____

⑮ Amy and Sara are best friends. They are 7 years old. How old will they be in 2 years? _____

Game Play the **Multiplication Table Game.**

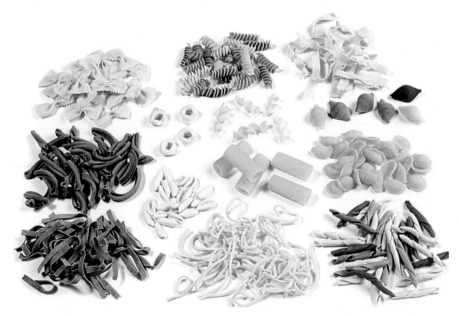

LESSON 11.6

Applying Multiplication

Key Ideas

Multiplication can be used to solve everyday problems.

Six slices of bread were used to make this sandwich.

How many slices of bread will be used to make 2 sandwiches?
$2 \times 6 = 12$

Solve these problems.

There are 5 vegetables on each plate.

1 How many vegetables are on 2 plates? _____

2 How many vegetables are on 3 plates? _____

3 How many vegetables are on 4 plates? _____

4 How many vegetables are on 5 plates? _____

One quiche costs $4.

5 How much do 6 quiches cost? _____

6 How much do 4 quiches cost? _____

Solve these problems.

7 CDs are sold in packages of 5. How many CDs are in 7 packages? _____

8 Charlie walks 4 blocks to school every day and walks home the same way. How many blocks does Charlie walk to and from school in 5 days? _____

9 Sally works 2 hours every morning and 3 hours every afternoon. How many hours does she work in 10 days? _____

10 Julia and Finn rode their bicycles 2 kilometers to the library. Each of them checked out 3 books. If they returned home the same way, how many kilometers was the round-trip? _____

11 Harriet bought 7 bottles of juice. Each bottle contained 8 ounces of juice and cost $3.

How many ounces of juice did she buy? _____
How much did the juice cost? _____

12 In football a touchdown is worth 6 points.
How many points are 6 touchdowns? _____

13 In basketball some baskets are worth 1 point, some baskets are worth 2 points, and some are worth 3 points.

Katie made five 3-point baskets and six 2-point baskets. How many points did she score? _____
Amy made two 3-point baskets and seven 2-point baskets. How many points did she score? _____

14 It costs $3 to rent a bike for one hour.
How much will a 7-hour rental cost? _____

 Game Play the **Multiplication Table Game.**

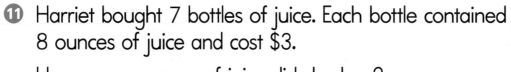

Name _____ Date _____

Listen to the problem.

Cathy is trying to find the total cost. She is making a table and using guess, check, and revise.

Number of Large Cans	Number of Small Cans	Total Number of Dolmades	Total Cost
2	4	$2 \times 8 = 16$ $4 \times 3 = 12$ $16 + 12 = 28$	
3			

Antonio is using logical reasoning
to find the total cost.

Solve the problem.

They should buy _____ large cans and _____ small.

They will pay _____ .

This is how I know.

Cumulative Review

Name _____ Date _____

Elapsed Time Lesson 10.10

Solve.

1 My family is making a South African dish. We made a marinade on Tuesday. It will be ready in three days. What day of the week will that be? _____

2 Mr. Liska is making sauerkraut on Sunday. He says good sauerkraut takes three weeks and three days to be perfect. What day of the week will it be when Mr. Liska finishes the sauerkraut? _____

Subtracting Two-Digit Numbers Lesson 6.4

Subtract.

3 81 − 68 = ____   **4** 100 − 25 = ____   **5** 19 − 19 = ____

6 92 − 45 = ____   **7** 33 − 11 = ____   **8** 43 − 37 = ____

Parts of Plane Figures Lesson 8.2

Ring the letter of the fraction shown.

9
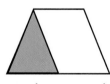

a. $\frac{1}{3}$ b. $\frac{1}{2}$ c. $\frac{1}{4}$ d. $\frac{1}{6}$

10

a. $\frac{1}{3}$ b. $\frac{1}{2}$ c. $\frac{1}{4}$ d. $\frac{1}{6}$

Cumulative Review

Shortcuts for Three-Digit Addition and Subtraction Lesson 9.10

Add or subtract.

⑪ $299 + 400 =$ _____

⑫ $400 - 100 =$ _____

⑬ $499 + 201 =$ _____

⑭ $401 - 99 =$ _____

Comparing Numbers through 1,000 Lesson 9.3

Draw the correct sign.

⑮ $100 \bigcirc 1{,}000$

⑯ $120 - 30 \bigcirc 80 + 10$

⑰ $230 + 40 \bigcirc 50 + 240$

⑱ $370 - 30 \bigcirc 360 - 30$

⑲ $18 \bigcirc 60 - 42$

⑳ $48 - 9 \bigcirc 39$

Using the Multiplication Table Lesson 11.5

Use the Multiplication Table to solve.

㉑ $4 \times 4 =$ _____

㉒ $5 \times 5 =$ _____

㉓ _____ $\times 3 = 9$

㉔ $3 \times 5 =$ _____

㉕ $2 \times 3 =$ _____

㉖ _____ $\times 4 = 8$

㉗ _____ $\times 2 = 2$

㉘ $0 \times 3 =$ _____

㉙ _____ $\times 5 = 0$

✕	0	1	2	3	4	5
0	0	0	0	0	0	0
1	0	1	2	3	4	5
2	0	2	4	6	8	10
3	0	3	6	9	12	15
4	0	4	8	12	16	20
5	0	5	10	15	20	25

LESSON 11.7 Missing Factors

Key Ideas

Missing-factor problems are similiar to missing-addend problems.

$$4 + \underline{\hspace{2cm}} = 7$$

$$4 \times \underline{\hspace{2cm}} = 12$$

Solve these problems. Use materials to help you solve.

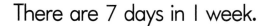

1 I paid $21 for 7 candles.

How much would 1 candle cost? _____

There are 7 days in 1 week.

2 How many weeks are in 14 days? _____

3 How many weeks are in 35 days? _____

4 How many weeks are in 49 days? _____

5 How many weeks are in 28 days? _____

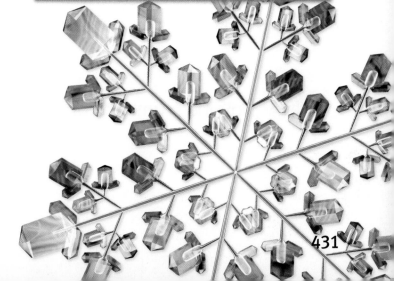

January						
SUN	MON	TUE	WED	THU	FRI	SAT
1	2	3	4	5	6	7
8	9	10	11	12	13	14
15	16	17	18	19	20	21
22	23	24	25	26	27	28
29	30	31				

e Textbook This lesson is available in the **eTextbook**.

Paul sells blueberries for $3 a basket.

Help Paul by completing this table.

Number of Baskets	1	2	3	4	5	6	7	8	9	10
Price (dollars)	3	6	9	12						

Use the table to answer these questions.

6 Megan paid $24. How many baskets did she buy?

$$\boxed{} \times 3 = 24$$

7 Late in the season Paul raised his prices. Five baskets of blueberries cost $20. How much was 1 basket? _____

$$5 \times \boxed{} = 20$$

Solve.

8 $4 \times \boxed{} = 8$

10 $5 \times \boxed{} = 20$

9 $5 \times \boxed{} = 15$

11 $10 \times \boxed{} = 50$

12 Nancy wants to buy 15 eggs. They come 5 to a box. How many boxes must she buy? $5 \times \boxed{} = 15$

Game Play the **Numbers on the Back Game.**

LESSON 11.8 Division and Multiplication

Key Ideas

Division is used to share parts of a whole.

I have 8 fortune cookies to share with 4 friends.
How many cookies will each friend get? 2

$8 \div 4 = 2$

Division and multiplication are inverse operations.
This is the symbol for division: $\div$.

Solve.

1 Twelve dollars are to be divided equally among
3 students. How many dollars will each student get?

$12 \div 3 = $ _____

2 Twenty-four dollars are to be divided equally among
4 students. How many dollars will each student get?

$24 \div 4 = $ _____

3 Thirty-two dollars are to be divided equally among
8 students. How many dollars will each student get?

$32 \div 8 = $ _____

4 Fifteen cents are to be divided equally among 3 cups.
How many cents will be in each cup?

$15 \div 3 = $ _____

Solve these problems. Use counters and a Multiplication Table if you need help.

⑤ 50 ÷ 10 = _____   **⑪** 40 ÷ 5 = _____

⑥ 5 × 10 = _____   **⑫** 8 × 5 = _____

⑦ 60 ÷ 6 = _____   **⑬** 40 ÷ 8 = _____

⑧ 10 × 6 = _____   **⑭** 5 × 8 = _____

⑨ 20 ÷ 5 = _____   **⑮** 48 ÷ 8 = _____

⑩ 4 × 5 = _____   **⑯** 6 × 8 = _____

Rich and Gail each need 40 stickers for a project.
There are 8 stickers in a package.
How many packages should they each buy?

Gail's Solution   Rich's solution

⑰ 8 × _____ = 40   **⑱** 40 ÷ 8 = _____

⑲ How many packages of stickers should Gail
and Rich each buy? _____

⑳ ▸Extended Response▸ How would you have solved the problem?

_____

_____

_____

  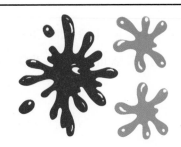

Name _____ Date _____

Division and Multiplication Functions

Key Ideas

Function tables can be used to practice multiplication and division as well as addition and subtraction.

Find the function rules.

1

in	out
5	15
16	26
27	37
49	59

The rule is _____.

3

in	out
1	3
2	6
3	9
4	12

The rule is _____.

2

in	out
29	29
31	31
45	45
72	72

The rule is _____.

4

in	out
8	4
6	3
10	5
12	6

The rule is _____.

Find the function rules. Fill in the missing numbers.

5

in	out
5	10
6	
8	
10	20

The rule is _____.

6

in	out
3	11
10	
20	28
	44

The rule is _____.

Create function-rule problems.
Challenge a friend to solve them.

7

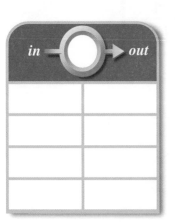

The rule is _____.

8

The rule is _____.

Name _____ Date _____

Listen to the problem.

8 Puppodums   8 Puppodums   8 Puppodums

Each person will get _____ puppodums.

Show how you know.

How many will each person get?

Each person will get _____ puppodums.

Show how you know.

Cumulative Review

Name _____ Date _____

Writing Fractions Lesson 7.2

Use the picture to answer the questions.

Jana and Min are riding the high-wire bikes at the museum. The wire is 20 meters long.

1 Jana has gone $\frac{3}{4}$ of the way. How many meters has she gone? _____

2 Min has gone $\frac{1}{4}$ of the way. How many meters has she gone? _____

3 Draw a J to show where Jana is.

4 Draw an M to show where Min is.

5 How many more meters does Min have to ride to reach the landing platform? _____

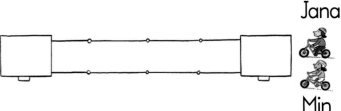

Jana

Min

Graphs on a Grid Lesson 4.12

Use the graph to answer the questions.

This restaurant has food from Central America. Use the graph to see how many customers have visited each day.

6 Connect the points on the graph.

7 About how many customers visited on day 1? _____

8 Which day had about 100 customers? _____

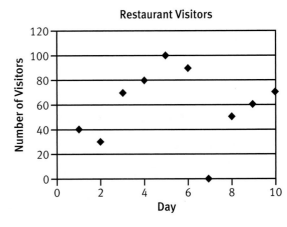

Restaurant Visitors

Cumulative Review

Division and Multiplication **Lesson 11.8**

Fill in the table. Then use the table to answer these questions.

9 Maya sells papayas for $4 a bag. She made this table to help her know how much to collect. Help Maya by completing the table.

Number of bags	1	2	3	4	5	6	7	8	9	10
Price (dollars)		8	12		20			32		

10 Ling bought 4 bags. What did she pay? $4 \times 4 = \boxed{}$ dollars

11 Niko paid $32. How many bags did he buy? $\boxed{} \times 4 = 32$

12 Miguel paid $40. How many bags did he buy? $\boxed{} \times 4 = 40$

13 Late in the season Maya raised her prices. Then she sold 3 bags of papayas for $18. What was the new price? _____ $3 \times \boxed{} = 18$

Name _____ Date _____

In this chapter you learned about multiplication and division. You learned to use multiplication to find the area of a figure. You learned to use division to solve problems about sharing.

Ring the correct answer.

1

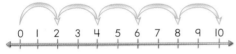

0 1 2 3 4 5 6 7 8 9 10

This number line represents skip counting by

a. 2.   **b.** 3.
c. 4.   **d.** 5.

2 Another way to write
$3 + 3 + 3 + 3$ is

a. $3 \times 3 \times 3 \times 3$.
b. 4×3.
c. 3×3.
d. $3 - 3 - 3 - 3$.

3

Rachel has 6 nickels. She wants to find out how many cents she has. Which is NOT a way she can find out?

a. 6×5
c. $5 + 5 + 5 + 5 + 5 + 5$

b. 5×5
d. Skip count by 5: *5, 10, 15, 20, 25, 30*

Use the Multiplication Table to solve the following.

4 $3 \times 4 =$ _____

5 $5 \times$ _____ $= 35$

✗	0	1	2	3	4	5	6	7	8	9	10
0	0	0	0	0	0	0	0	0	0	0	0
1	0	1	2	3	4	5	6	7	8	9	10
2	0	2	4	6	8	10	12	14	16	18	20
3	0	3	6	9	12	15	18	21	24	27	30
4	0	4	8	12	16	20	24	28	32	36	40
5	0	5	10	15	20	25	30	35	40	45	50
6	0	6	12	18	24	30	36	42	48	54	60
7	0	7	14	21	28	35	42	49	56	63	70
8	0	8	16	24	32	40	48	56	64	72	80
9	0	9	18	27	36	45	54	63	72	81	90
10	0	10	20	30	40	50	60	70	80	90	100

Write your answers.

6 The area is _____ square units.

7 **Extended Response** Explain how you found the area.

_____

_____

8 There are 3 friends who want to share
9 fortune cookies. Write the division problem
that shows how many fortune cookies each
friend will get.

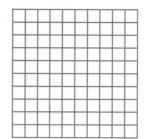

_____ ÷ _____ = _____

Find the function rules, and fill in the
missing numbers.

9

in → out	
5	10
6	
7	
10	20

The rule is _____.

10

in → out	
8	4
6	
	5
12	6

The rule is _____.

Name _____ Date _____

Lesson 11.1 **Skip** count and fill in the missing numbers.

1 0   6   12   18   _____   30

2

2	4	6		10		14			20

Lesson 11.9 **Find** the function rules. Fill in the missing numbers.

3

in	out
8	4
12	6
16	8
22	11

The rule is _____.

5

in	out
8	24
9	
10	
11	33

The rule is _____.

4

in	out
98	93
107	102
83	78
62	57

The rule is _____.

6

in	out
18	6
9	
12	
24	8

The rule is _____.

Lesson 11.5 **Use** the squares to solve these problems.

7 The area is _____ square units.

8 $4 \times 13 = $ _____

Lesson 11.7 **Solve** these problems.

9 I paid $21 for seven candles.
How much would one candle cost? $_____

10 $54 is to be divided equally among nine people.
How many dollars for each person?

$54 \div 9 = $ _____

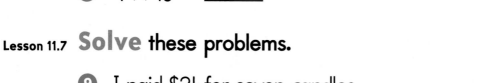

Lessons 11.2, 11.6 **Solve.**

Each plate has four tamales.

11 How many tamales are there? _____

12 $5 \times \boxed{} = 15$

14 $5 \times \boxed{} = 20$

13 $\boxed{} \times 4 = 8$

15 $4 \times \boxed{} = 8$

16 Mira wants to make a Greek salad for 35 people.
Her recipe makes enough for 7 people. How many
times will she use her recipe?

$7 \times \boxed{} = 35$

Name _____ Date _____

Skip count to find the missing numbers.

1 3, 6, 9, 12, _____, _____, _____, 24

2 4, 8, 12, 16, _____, _____, _____, 32

3 5, 10, 15, 20, _____, _____, _____, 40

Solve.

4 $8 + 8 + 8 =$ _____

How many times did you add 8? _____ times

$3 \times 8 =$ _____

5 $7 + 7 + 7 + 7 + 7 =$ _____

How many times did you add 7? _____ times

$5 \times 7 =$ _____

Use pictures to solve.

6 $4 + 4 + 4 =$ _____

$3 \times 4 =$ _____

7 $9 + 9 + 9 =$ _____

$3 \times 9 =$ _____

8 $5 + 5 + 5 + 5 + 5 =$ _____

$5 \times 5 =$ _____

Multiply.

9 1 × 6

a. 6   b. 7   c. 10   d. 16

10 7 × 7

a. 35   b. 42   c. 49   d. 56

11 9 × 5

a. 50   b. 45   c. 40   d. 35

12 4 × 3

a. 7   b. 12   c. 16   d. 43

Multiply or divide.

13 2 × _____ = 12

a. 4   b. 5   c. 6   d. 7

14 _____ × 5 = 20

a. 15   b. 6   c. 5   d. 4

15 16 ÷ 4 = _____

a. 12   b. 8   c. 6   d. 4

16 42 ÷ 7 = _____

a. 6   b. 7   c. 8   d. 9

Name _____ Date _____

Find the rule for each function table.

⑰

in	out
8	4
6	2
10	6
12	8

a. +4   **c.** ×2
b. −4   **d.** ÷2

⑱

in	out
4	8
6	12
5	10
3	6

a. ×2   **c.** +5
b. +4   **d.** ×3

Ring the time.

⑲ Kitri started her school day at 8:00 A.M. School was out for the day 7 hours later. What time was it when Kitri got out of school?

a. 1:00 A.M.   **b.** 2:00 P.M.   **c.** 3:00 A.M.   **d.** 3:00 P.M.

⑳ Russell opened his store at 10:00 A.M. He closed the store and went home at 9:00 P.M. How many hours was the store open?

a. 12 hours   **b.** 11 hours   **c.** 9 hours   **d.** 1 hour

㉑ Tyler and Landon went to the carnival at 11:00 A.M. on Friday. They stayed until 4:00 P.M. How many hours did they spend at the carnival?

a. 4 hours   **b.** 5 hours   **c.** 6 hours   **d.** 7 hours

Draw a picture to solve.

22 Ricardo drew 2 rectangles on grid paper. Each rectangle had an area of 18 square units, but they did not have the same length. The length and the width were more than 1 square long. Draw Ricardo's rectangles.

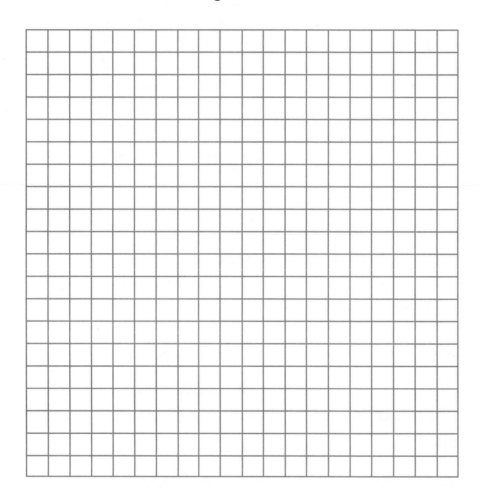

23 Choose one of the rectangles that you drew. Write two multiplication facts and two division facts that the rectangle shows.

The Amazing Number Machine
Makes Mistakes

Draw your own number machine.
Write what it does.

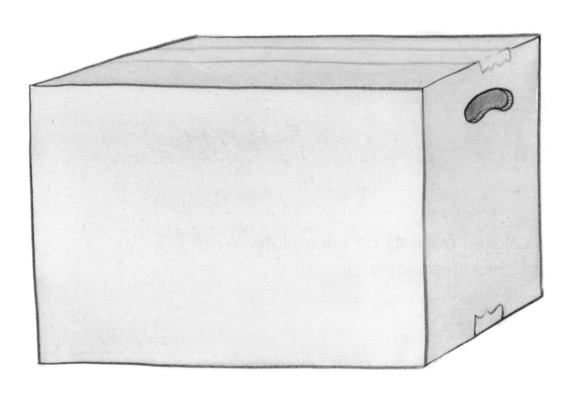

_____

_____

_____

_____

Count how many people will be at the party.
Write the number. _____

Draw 3 chicken nuggets on each plate. Write the
number of chicken nuggets altogether. _____

In This Chapter You Will Learn
- how to create and extend patterns.
- how to create word problems.

451

Name _____ Date _____

Listen to the problem.
Draw your pattern on the grid.

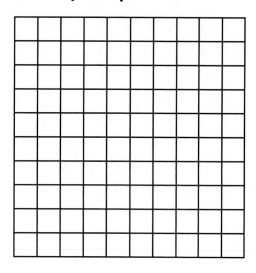

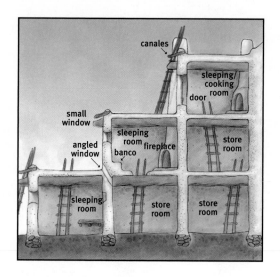

Describe your pattern so someone else
can make it.

_____

_____

_____

_____

_____

_____

Patterns and Functions

Key Ideas

Patterns occur often in the world around us. When you recognize a pattern, you can figure out what comes next.

Find the patterns. Fill in the blanks.

① 30, 27, 24, 21, 18, 15, _____, _____, _____, _____, 0

② 100, 95, 90, 85, _____, _____, _____,

_____, _____, _____, 50

③ 7, 14, 21, 28, _____, _____, _____, _____, _____, 70

④ 70, 63, 56, 49, _____, _____, _____, _____,

_____, _____, 0

⑤ 1, 3, 5, 4, 6, 8, 7, 9, _____, _____, _____,

_____, _____, 15

⑥ 3, 6, 9, 7, 10, 13, 11, _____, _____, _____, _____, 21

⑦ 2, 4, 6, 5, 7, 9, 8, 10, _____, _____, _____,

_____, _____, 16

⑧ 20, 17, 14, 16, 13, 10, 12, _____, _____, _____, _____, 2

Fill in the blanks.

9

in → +4 → out	
3	7
10	
0	4
	24

10

in → −5 → out	
10	5
14	
23	
	0

What is the rule? Write it in the box.

11

in → ○ → out	
7	9
10	12
81	83

The rule is ☐ .

12

in → ○ → out	
70	60
40	30
27	17

The rule is ☐ .

Writing + Math **Journal**

How are patterns useful when you are trying to solve a problem?

LESSON 12.2 Functions with Mixed Operations

Key Ideas

A mixed operation is an expression that uses more than one operation.

$3 + 8 - 4$ is a mixed operation because it uses addition and subtraction.

$3 + 8 - 4 = 7$

Follow the directions.

1 Kevin is making hard function problems. His rule is $+9 -4$. Fill in the missing numbers.

2 Write what is difficult about Kevin's problem.

_____

3 Write what is easy about Kevin's problem.

_____

8	13
20	
3	
10	15

4 Lorena is also making hard function problems. Her rule is $\times 3 -1$. Fill in the missing numbers.

5 Write what is hard about Lorena's problem.

_____

6 Write what is easy about Lorena's problem.

_____

3	8
10	
4	
2	5

Find the mixed operation function.

7

4	7
5	9
6	11
7	13

The rule is _____.

8

6	13
7	15
8	17
9	19

The rule is _____.

Make your own mixed operation functions. Challenge a friend to solve them.

9

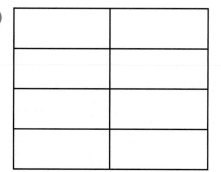

The rule is _____.

10

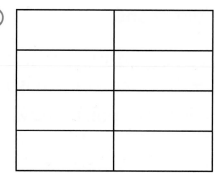

The rule is _____.

11

The rule is _____.

12

The rule is _____.

LESSON 12.3

More Functions with Mixed Operations

Key Ideas

Mixed operations can help you solve everyday problems quickly.

Matt charges $3 per hour for mowing lawns, plus $1 to cover the cost of his travel.

1. Complete the table to show how much Matt should collect.

Matt's Lawn Mowing Service Price Table

Hours	1	2	3	4	5	6	7	8
Charge (dollars)	4	7	10					

Allison also mows lawns. She estimates how long a job will take. Then she charges a fixed price for her service.

2. **Extended Response** Which method of charging is fairer, Matt's or Allison's? Write why you think so.

_____

_____

_____

Kenji is busy making hard function problems.

Complete the tables, and then find an easier rule to see why they are not really that hard. Then write the simplified rule for each problem.

+5 −2

1	4
0	3
10	13

③ The simplified rule is _____.

+5 +5

1	11
0	10
18	28

⑤ The simplified rule is _____.

×2 ÷2

8	8
1	1
17	17

④ The simplified rule is _____.

×6 ÷2

1	3
0	0
7	21

⑥ The simplified rule is _____.

Name _____ Date _____

Patterns and Shapes

Key Ideas

Plane figures can be used to make a pattern.

Look for the pattern. Draw the missing figures.

1

2

3

4

5

6

Make your own patterns using colors.

7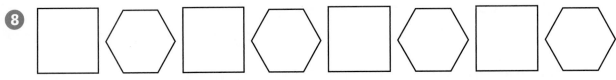

8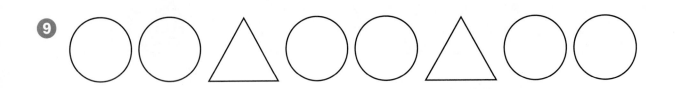

9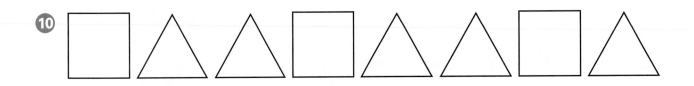

10

11

12

Name _____ Date _____

Listen to the problem.

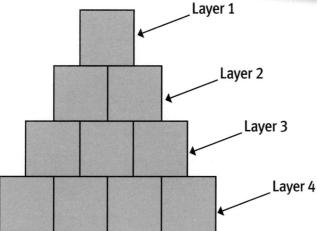

Layer 1

Layer 2

Layer 3

Layer 4

Mark made a physical model to solve the problem.

Rebecca used draw a diagram/use a number pattern to solve the problem.

 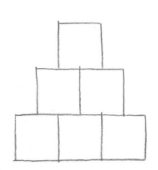 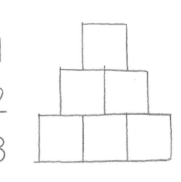

1
2
3
4
5
6

Draw a picture of how you solved the problem.

A model of the wall with 10 layers would

need _____ blocks.

Cumulative Review

Name _____ Date _____

Copyright © SRA/McGraw-Hill.

Applying Multiplication Lesson 11.6

Solve these problems.

① Malin wants to buy 36 cups of salt. There are 4 cups of salt in each bag. How many bags must she buy?

$4 \times \boxed{} = 36$

② Jillian made 96 ounces of jam. She wants to store the jam in jars. She can fit 12 ounces in each jar. How many jars will she need?

$12 \times \boxed{} = 96$

Checking Subtraction Lesson 6.7

Solve these problems. Explain your answers.

③ Ellie had $83. She bought a kite. She now has _____.

Explain. _____

④ Aiden bought a basketball. He gave the clerk two $20 bills. How much change did he get? _____

Explain. _____

⑤ Daniel has a $50 bill. He buys two dinosaur models and a kite. How much change should he get? _____

Explain. _____

Cumulative Review

Copyright © SRA/McGraw-Hill.

Patterns and Functions Lesson 12.1

Find the patterns. Fill in the blanks.

6

2	4	6			14	16	18			24

7

3	6	9	12	15	18	21					36

8

in → +5 → out	
4	9
53	
9	14
	32

9

in → −11 → out	
60	49
	11
30	19
	7

Obtuse, Acute, and Right Angles Lesson 8.5

Write the time. Then write whether the angle is obtuse, acute, or right.

10  **11**  **12**  **13**

_____   _____   _____   _____

_____   _____   _____   _____

14 Ring the right angles.

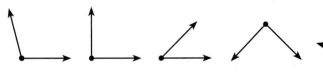

Real Math • Chapter 12

LESSON 12.5

Column Addition

Key Ideas

You can add several large numbers together by using strategies and shortcuts you already know for adding one- or two-digit numbers.

Add.

1
```
   342
   186
 + 274
```

2
```
   98
   76
 + 85
```

3
```
   343
   297
 + 160
```

4
```
   333
   333
 + 334
```

5
```
   75
   75
   75
 + 75
```

6
```
   250
   250
   250
 + 250
```

7
```
   176
   321
   219
 + 134
```

8
```
   54
   87
   69
 + 94
```

9
```
   576
    89
   112
 + 96
```

 Textbook This lesson is available in the *eTextbook*.

Answer the questions.

10 Mr. Li is updating his horse corral. He wants to put a fence around it. What is the perimeter of the corral he wants to fence?

_____

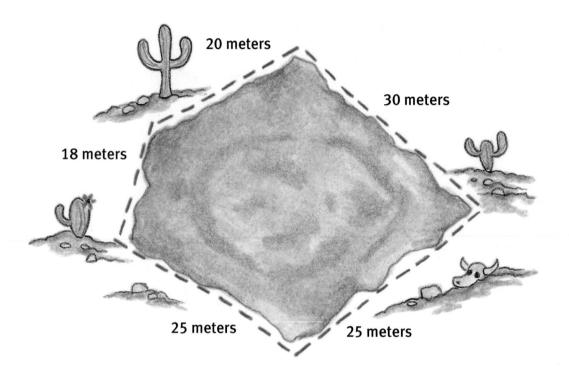

20 meters

30 meters

18 meters

25 meters   25 meters

11 Each roll of fencing is 25 meters long. How many rolls of fencing does Mr. Li need? _____

12 The town of Muddleville was founded 96 years ago. There are 87 people now living in Muddleville. There are also 42 cats in Muddleville. How many dogs are there in Muddleville?

_____

 Play the **Roll a Problem Game.**

LESSON 12.6

Number Sentences

Key Ideas

There are many different ways to combine numbers and operations to represent a given number.

Write three number sentences that give the number shown as the answer.

10
1 _____
2 _____
3 _____

8
4 _____
5 _____
6 _____

5
7 _____
8 _____
9 _____

Write a number sentence for each situation. Then solve the problem.

10 Mary earned $5 on Monday, $7 on Tuesday, and $4 on Wednesday. How many dollars did she earn altogether?

_____

11 Samantha had 9 marbles. She bought 2 more. Then she gave away 3. Then she got 3 more. How many marbles does Samantha have now?

_____

12 John earned $3 a day for 7 days. How many dollars has he earned altogether?

_____

13 There were 20 apples on the tree. Abby picked 3 apples a day for 5 days. One apple fell off the tree. How many apples are on the tree now?

_____

 Play the **Harder What's the Problem Game.**

LESSON 12.7 Creating Word Problems

Key Ideas

To show a missing number in a problem, you can use a **variable**. A variable is a symbol that stands for a number you do not yet know.

Follow the directions for each problem.

Monica invited 12 people to her party. Some of the people didn't come.

1 Write a number sentence to show how many people came. Use x to stand for the number of people who didn't come and y to stand for the number who came.

_____

2 If 4 didn't come, how many came? _____

Danny is bringing a snack to school. He wants to make sure he brings 2 napkins for each student in his class plus 10 extra napkins in case there is a spill.

3 Write a number sentence to show how many napkins Danny needs to bring. Use x to stand for the number of students and y to stand for the number of napkins he needs to bring.

_____

4 If 20 students are in his class, how many napkins will Danny need to bring? _____

e Textbook This lesson is available in the *eTextbook*.

Write word problems for each number sentence below.

5 $y = 5 + x$

_____

_____

_____

6 $y = 10 - x$

_____

_____

_____

Name _____ Date _____

Listen to the problem.

Draw your rug pattern.

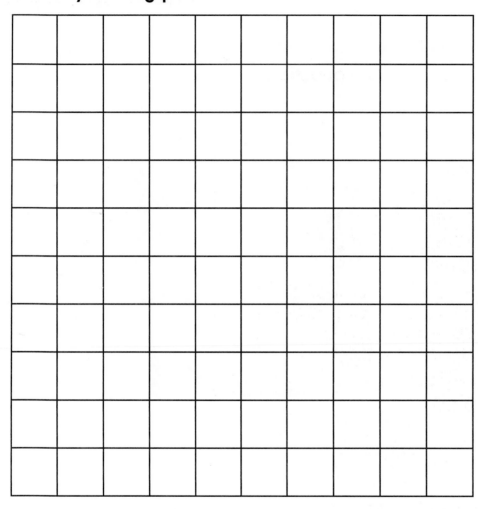

Describe your pattern so that someone else can make the same rug.

_____

_____

_____

_____

_____

Cumulative Review

Copyright © SRA/McGraw-Hill.

Name _____ Date _____

Horizontal Bar Graphs Lesson 4.11

Use the graph to answer the questions.

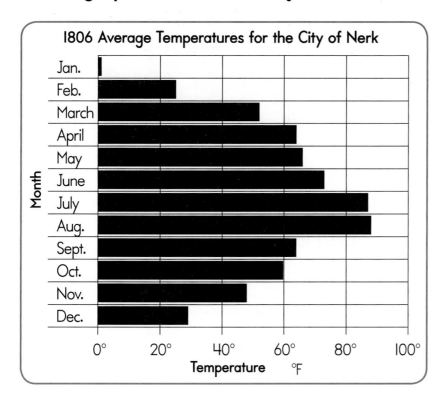

1806 Average Temperatures for the City of Nerk

1. What was the average temperature for October? _____ °F

2. What was the average temperature for March? _____ °F

3. What month was the coldest in this city? _____

4. What month was the warmest? _____

Cumulative Review

Column Addition **Lesson 12.5**

Use the drawing to answer the questions.

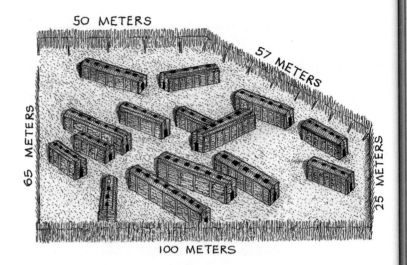

5 This is a village with fifteen Iroquoian longhouses. There is a fence around the longhouses. What is the perimeter of the fence?

_____

6 It took 1 bundle of sticks to make 10 meters of the fence. How many bundles of sticks did they need?

_____

Number Sentences **Lesson 12.6**

Write three number sentences that give the number shown as the answer.

14 7 _____ 8 _____ 9 _____

24 10 _____ 11 _____ 12 _____

36 13 _____ 14 _____ 15 _____

Name _____ Date _____

In this chapter you explored patterns and used numbers in different ways. You learned to solve problems with mixed operations. You learned to use variables.

....

Find the patterns. Fill in the blanks.

1 100, 95, 90, 85, _____, _____, _____, 65

2 7, 14, 21, 28, _____, _____, _____, 56

3 ▢ ▲ ▢ ____ ____ ____ ▲ ▢

Follow the directions.

4 Kayla made a function problem. Her rule was +8 −4. Rewrite the rule for Kayla's function machine using only one operation. _____

5 James earns $10 every afternoon he helps at his uncle's fruit stand. However, it costs James $1 to ride the bus to the stand and $1 to ride the bus home. Write the number sentence that shows how many dollars James earns each afternoon he helps his uncle.

_____

Write your answer.

6 There are 87 students in second grade, 92 in third, 78 in fourth, and 81 in fifth. How many students are there? _____

Fill in the blanks with numbers to make each number sentence true.

7 _____ + _____ + _____ − _____ = 10

8 _____ − _____ − _____ = 8

Write a number sentence to represent each problem. Use variables to represent the numbers you do not know.

9 Marco read 6 books in June, 8 books in July, and a few more books in August. How many books did Marco read during the summer?

_____

10 Ana picked 3 apples per day for 6 days. How many apples did Ana pick?

_____

Name _____ Date _____

Lessons 12.1, 12.4 Look for the pattern. Then draw the missing shape or missing number.

❶

❷

❸ 42, 39, 36, 33, 30, 27, _____, _____, _____, _____, 12

❹ 96, 108, 120, 132, _____, _____, _____, _____, 192

❺ 18, 22, 26, 30, _____, _____, _____, _____, _____, 54

❻ 82, 73, 64, 55, 46, _____, _____, _____, _____, 1

Lesson 12.2 Fill in the missing numbers.

❼

+8	−4
9	13
18	
13	
20	24

❽

×4	−2
2	6
5	
3	
6	22

Lesson 12.5 ## Solve.

⑨	568	⑩	46	⑪	96	⑫	555
	423		782		521		777
	+ 167		+ 31		+ 362		+ 622

Lesson 12.6 ## Write a number sentence for each situation. Then solve the problem.

⑬ Lin earned $40 in January, $30 in February, and $20 in March. How many dollars did she earn in the three months? _____

⑭ Lilly made 11 biscuits. Her brother ate 5 biscuits. She made 8 more. Then her brother ate 4 more. How many biscuits does Lilly have now? _____

⑮ Evan worked 8 hours a day for 6 days. How many hours has he worked? _____

Lesson 12.3 ## Write the simplified rule for each problem.

+8 −3		×5 ÷5		+8 +7	
18	23	12	12	4	19
7	12	15	15	14	29
21	26	20	20	27	42

⑯ The simplified rule is _____.

⑰ The simplified rule is _____.

⑱ The simplified rule is _____.

Name _____ Date _____

Find the pattern. Fill in the blanks.

1 45, 40, 35, 30, _____, _____, 15

The pattern is _____.

2 9, 12, 15, 18, _____, _____, 27

The pattern is _____.

3 2, 4, 6, 8, _____, _____, 14

The pattern is _____.

4 10, 20, 30, 40, _____, _____, 70

The pattern is _____.

Complete the table to solve.

5 Mr. Beck had a coupon to buy green beans. One case of beans cost $6. Every additional case was $4 each. Find out how much he would spend to buy 2, 3, 4, 5, and 6 cases.

Green Beans with Coupon						
Cases	1	2	3	4	5	6
Cost $	6					

6 A store had a sale on T-shirts. The first shirt cost $10, and every additional shirt was $7. Find the cost of 2, 3, 4, 5, and 6 shirts.

T-Shirt Sale						
Shirts	1	2	3	4	5	6
Cost $	10					

Find the mixed operation rule.

7

in	out
3	7
6	16
10	28
5	13

a. × 3 − 2   **b.** × 3 + 2
c. × 4 − 1   **d.** × 4 + 1

8

in	out
20	11
14	8
10	6
12	7

a. × 2 + 1   **b.** × 2 − 1
c. ÷ 2 − 1   **d.** ÷ 2 + 1

Ring the letter of the number sentence that matches the problem.

9 Gracie earned $5 per hour for 6 hours. She put this money in the bank, where she already had $20.

a. 20 ÷ 5 + 6   **b.** 20 ÷ 6 + 5
c. 20 × 6 + 5   **d.** 5 × 6 + 20

10 Kip baked 7 trays of cookies. There were 8 cookies on each tray. Then he ate 3 of the cookies.

a. 7 × 3 − 8   **b.** 7 × 8 − 3
c. 8 × 3 − 7   **d.** 8 + 7 − 3

Name _____ Date _____

Add.

⑪   214
    + 507

   a. 293
   b. 311
   c. 711
   d. 721

⑫   726
    + 198

   a. 924
   b. 904
   c. 814
   d. 528

⑬   234
     157
    + 108

   a. 342
   b. 391
   c. 499
   d. 501

⑭   330
     175
    + 119

   a. 624
   b. 505
   c. 449
   d. 294

Solve.

⑮ $\frac{1}{3}$ of 30

   a. 15
   b. 10
   c. 9
   d. 3

⑯ $\frac{1}{4}$ of 20

   a. 20
   b. 15
   c. 10
   d. 5

⑰ $\frac{3}{3}$ of 30

   a. 33
   b. 30
   c. 15
   d. 10

⑱ $\frac{1}{2}$ of 60

   a. 10
   b. 15
   c. 30
   d. 60

⑲ A square has one side that is 3 feet long.
   What is the perimeter?

   a. 3 ft   **b.** 6 ft   **c.** 9 ft   **d.** 12 ft

⑳ A triangle has two sides that are 3 inches long and
   one side that is 5 inches long. What is the perimeter?

   a. 5 in.   **b.** 8 in.   **c.** 10 in.   **d.** 11 in.

㉑ 3 ft = _____

   a. 24 in.   **b.** 3 yd   **c.** 1 yd   **d.** 1 in.

Extended Response **Use** shapes to draw patterns.

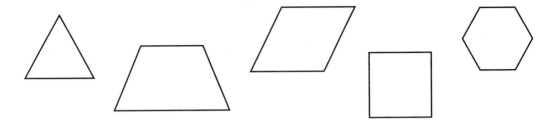

22 Place 2 hexagons and 4 squares in any order you choose to make a pattern. Draw the pattern you made. Draw lines to show each time the pattern repeats.

23 Chan used two different shapes to draw a pattern. The shape he used first was used more than the other shape. Draw a pattern that could be like the one Chan drew. Draw lines to show each time the pattern repeats.

Thinking Story

Paint Up, Fix Up, Measure Up

Each side of the house takes 10 gallons of paint. Mr. Muddle has leftover paint. Each bucket holds 5 gallons of paint. Write the name of the color Mr. Muddle will need to use to paint all four walls of his house.

_____

How many centimeters too wide is the screen
for the door on each side? Write the number.   _____

How many centimeters too short is the screen
for the door? Write the number.   _____

Real Math • Chapter 12

A

acute angle

angles

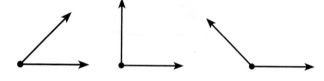

C

cylinder

E

eighth

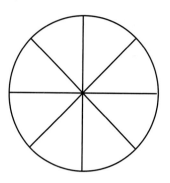

F

fifth

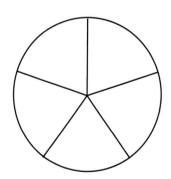

flip

Picture Glossary

M

mode

the number that appears most often in a set of data

O

obtuse angle

P

parallel

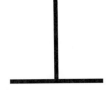

perpendicular

polygon

a closed plane figure with three or more sides

prism

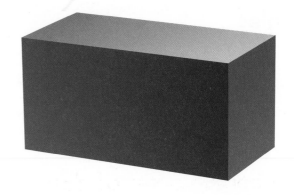

pyramid

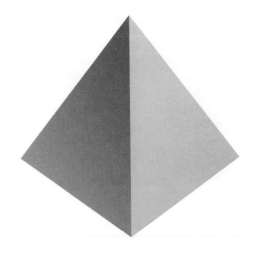

Q

quadrilateral

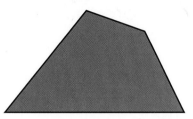

R

range

the lowest and highest numbers in a set of data

rhombus

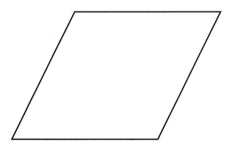

right angle

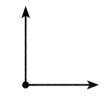

S

sixth

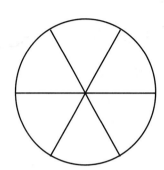

slide

Picture Glossary

T

thermometer

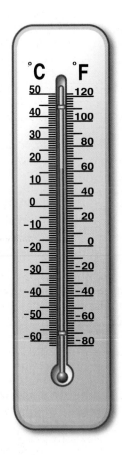

trapezoid

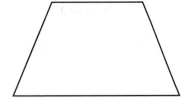

turn

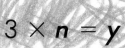

V

variables

$$3 \times n = y$$

Photo Credits

1, © Index Stock Imagery, Inc.; **3,** © Matt Meadows; **12,** © Matt Meadows; 20, © Comstock/Getty Images, Inc.; **22,** © Photodisc/ Getty Images, Inc.; **24,** © Dorling Kindersley/Getty Images, Inc.; **25,** © Corbis; **29–36,** © Matt Meadows; **41,** © Stone/Getty Images, Inc.; **44,** © Chris Butler/AgeFotostock/SuperStock; **45,** © Photodisc/Getty Images, Inc.; **46**(c), © Corbis, (bl), © Georgette Douwma/Science Photo Library/Photo Researchers, Inc., (br), © Stone/Getty Images, Inc.; **47,** © Photodisc/Getty Images, Inc.; **51,** © Photodisc/Getty Images, Inc.; **52,** © H. Spichtinger/zefa/Corbis; **58,** © Photodisc/Getty Images, Inc.; **63**(t), © Marevision/ AgeFotostock/SuperStock, (b), © Photodisc/Getty Images, Inc.; **64,** © Photodisc/Getty Images, Inc.; **79,** © Taxi/Getty Images, Inc.; **81,** © Melissa Lockhart/Superstock; **82,** © Steve Satushek/Getty Images, Inc.; **83,** © Matt Meadows; **84,** © Digital Vision Ltd./ Superstock; **92,** © Digital Vision Ltd./Superstock; **92,** © Matt Meadows; **93**(t), © Melissa Lockhart/Superstock, (b), © Digital Vision Ltd./Superstock; **94,** © Digital Vision Ltd./Superstock; **96,** © Lawrence Manning/Corbis; **113,** © Powered by Light/ Alan Spencer/Alamy Images; **115,** (t), (tc), (c), © Matt Meadows, (b), © Morton and White; **116,** © Matt Meadows; **117,** (t), (b), © Matt Meadows, (cl), © Photodisc/Getty Images, Inc., (cr), © Morton and White; **119,** © Photodisc/Getty Images, Inc.; **120**(t), © Stockdisc/Getty Images, Inc., (b), © CheapShots/Alamy; **121,** © Brand X Pictures/Getty Images, Inc.; **122,** © Photodisc/ Getty Images, Inc.; **123, 124,** © Stockdisc/Getty Images, Inc.; **127**(t), © Stockbyte, (c), © Photodisc/Getty Images, Inc., (b), © Photodisc/Getty Images, Inc.; **128,** © Matt Meadows; **133, 134,** © Photodisc/Getty Images, Inc.; **135,** © Matt Meadows; **136,** © Photodisc/Getty Images, Inc.; **139,** © Panoramic Images/Getty Images, Inc.; **140,** © Brand X Pictures/Getty Images, Inc.; **141,** © Stone/Getty Images, Inc.; **142,** © Photodisc/Getty Images, Inc.; **159,** © Photodisc/Getty Images, Inc.; **161,** © Matt Meadows; **163, 164,** © Photodisc/Getty Images, Inc.; **165,** © Matt Meadows; **166,** © Photodisc/Getty Images, Inc.; **168–175,** © Matt Meadows; **178,** © Photonica/Getty Images, Inc.; **181,** © Foodpix/Getty Images, Inc.; **183,** © Photographer's Choice/Getty Images, Inc.; **184,** © Morton and White; **185,** © Matt Meadows; **186,** © Index Stock Imagery, Inc.; **187,** © Deborah Davis/PhotoEdit; **204,** © Kevin Schafer/Corbis; **205,** © Photonica/Getty Images, Inc.; **206,** © The Image Bank/Getty Images, Inc.; **208,** © McDonald Wildlife Photography/Animals Animals-Earth Scenes; **210,** © Joseph Sohm/ChromoSohm Inc./Corbis; **211,** © Panoramic Images/Getty Images, Inc.; **212,** © Index Stock Imagery, Inc.; **222**(t), ©Stockdisc/Getty Images, Inc., (b), © Photodisc/Getty Images, Inc.; **225**(tl), © Dorling Kindersley/Getty Images, Inc., (tc), © Photodisc/Getty Images, Inc., (tr), © The Image Bank/Getty Images, Inc.; **226**(t), © Photodisc/Getty Images, Inc., (b), © Taxi/Getty Images, Inc.; **243,** © Michael Newman/PhotoEdit; **245, 246,** © Matt Meadows; **250,** © Brand X Pictures/Getty Images, Inc.; **254,** © Photodisc/Getty Images, Inc.; **255,** © Matt Meadows; **293,** © Matt Meadows; **300,** © Christian Darkin/Alamy; **304,** © Francois Gohier/Western Paleontological Labs/Photo Researchers, Inc.; **306,** © Jim Zuckerman/Corbis; **307,** © Manfred Kage/Peter Arnold, Inc.; **308,** © Elena Ray/Alamy; **323,** © Photodisc/Getty Images, Inc.; **329,** © Walter Chandoha/Chandoha Photography; **334,** © Dorling Kindersley/Getty Images, Inc.; **336,** © David Young-Wolff/ PhotoEdit; **337, 338** © Photodisc/Getty Images, Inc.; **340,** © Stockdisc/Getty Images, Inc. **342,** © Peter Mason/Getty Images, Inc.; **344,** © Index Stock Imagery, Inc.; **346,** © Photodisc/Getty Images, Inc.; **351,** © Matt Meadows; **352,** © K. Handke/zefa/Corbis; **355,** © Photodisc/Getty Images, Inc.; **356,** © Brand X Pictures/Getty Images, Inc.; **357,** ©Index Stock Imagery, Inc.; **358,** (t), © Photodisc/Getty Images, Inc., (cl), (c), (cr), © Matt Meadows; **378,** © Iconica/Getty Images, Inc.; **382**(t), © Photodisc/ Getty Images, Inc., (b), © Panoramic Images/Getty Images, Inc.; **384**(l), © Brand X Pictures/Getty Images, Inc., (r), © Photodisc/ Getty Images, Inc.; **389, 390,** © Matt Meadows; **391,** © Stockdisc/Getty Images, Inc.; **392**(t), © Stockdisc/Getty Images, Inc., (tc), © Corbis, (bc), © Photodisc/Getty Images, Inc., (b), © Corbis; **394,** © Stockdisc/Getty Images, Inc.; **396**(l), © Corbis, (r), © Brand X Pictures/Getty Images, Inc.; **398**(t), © Photodisc/Getty Images, Inc., (b), © Photographer's Choice/Getty Images, Inc.; **408,** © Photodisc/Getty Images, Inc.; **409,** © Brand X Pictures/Getty Images, Inc.; **413,** © Mark Baynes/Alamy Images; **416,** © Matt Meadows; **418,** © Foodpix/Getty Images, Inc.; **419**(t), © Foodpix/PictureArts Corporation, (cl), © Foodpix/PictureArts Corporation, (cr), © Photodisc/Getty Images, Inc., (bl), © Matt Meadows, (br), © Stockdisc/Getty Images, Inc.; **420**(t), (b), © Matt Meadows, (c), © BrandXPictures/Getty Images, Inc.; **421,** © Photodisc/Getty Images, Inc.; **422,** © Foodpix/PictureArts Corporation; **424**(t), © Photodisc/Getty Images, Inc., (b), © Alamy Images; **425**(t), © Foodpix/PictureArts Corporation, (c), (b), © Matt Meadows; **426,** © Corbis; **431,** © Guy Grenier/Masterfile; **432, 433** © Photodisc/Getty Images, Inc.; **434,** © Matt Meadows; **436,** © Photodisc/Getty Images, Inc.; **446,** © Brand X Pictures/Getty Images, Inc.; **453,** © Werner Forman/Corbis; **457,** © Brand X Pictures/Getty Images, Inc.; **465,** © Nik Wheeler/Corbis; **467,** © Iconica/Getty Images, Inc.; **469,** © Daved Muench/Corbis; **470,** © Tim Davis/Corbis.

Freddy

bonnie

chica